Probability for the Game Player
Another six, a head and an ace

BOOK TWO

Probability Devices, Influences and Game Play

John A. Bower

A continuing guide to exploring probability with dice, coins, cards and other randomising devices and influences in games and puzzles with spreadsheet calculation

Includes sections on games at home on the table top, board games, video versions in computers and tablets, casino games, lottery, dice, card and sport games

About the author

John Bower is a retired science lecturer (Queen Margaret University, Edinburgh), whose interest in probability stems from a developing study of statistical applications and research methods and an enduring fascination with games. His current works include the **Probability for the Game Player** series, where his statistical knowledge focuses on probability in the realm of game play. This publication (Book Two: Probability Devices, Influences and Game Play) is also available in a Kindle version. Book One in the series, (Probability Basics) is available in print and Kindle versions. He has written several journal articles and one textbook on the use and application of statistics in an applied field (food science & technology).

To Jonathan, Cassandra and to Pushka-Latitia with love and to all my family and friends who played games with me over the years. Special thanks to Cassie, Johnny and Nick for assistance with probability trials and calculations.

Table of Contents

Preface

BOOK TWO

This is the second book in the series ***Probability for the Game Player***, in which Book One introduced probability and laid some foundations. This volume can be viewed as an application book. It continues the exploration of probability in games and begins by setting the scene for game play for games, gambling and some sport game applications. In the former activities, randomizing devices decide events, whereas in the latter, small random influences exert an effect on outcomes. Following chapters look in more depth at the probabilities for various randomizing devices and their use in a selection of problems, puzzles and games, ranging from recreational play to casino gambling. These include reference to coins, dice, cards, roulette, lottery, board and role-play games, in their original versions and with some reference to online video counterparts. Chapter numbering follows on from Book One, i.e. chapters number from 15 to 22 and each can be read in isolation for the most part. 19th November 2014

BOOK ONE (separate volume)

Do you dabble in games, puzzles or sports statistics where the word 'probability' crops up? Or maybe you are doing a math or statistics module that has probability content and you find that you have difficulty in understanding some sections? If so, then you may find some interesting things in the pages that follow. This book will provide you with a different viewpoint to that of the purely mathematical. It presents probability as a subject for study, interest and enjoyment in the context of games of several kinds, with the advantage of more explanation and different, more understandable views of some aspects. My interest in probability stems from a developing interest in statistical applications and research methods as a science lecturer (retired) and an enduring fascination with games.

29th April 2014

Trademark notice

The following are trademarks or registered trademarks of their
respective companies:

Excel is a trademark of Microsoft ® Corporation

Chapter 15
Games on the table top, in casinos, computers and in sport

15.1 Introduction

The first chapter of Book Two looks at probability from a number of viewpoints, illustrated with a variety of random events. Probabilities for single and multiple devices are included, covering the various formats of events (sequence, content, run, etc.), with formulae, functions and rules from Book One. These are referred to in full and serve to refresh the core material for later chapters on specific devices. The main theme of exploration is continued, namely, probabilities are calculated and established. Wherever possible, they are used to give improved game play and strategy, but this does not apply in every case.

First, the perception of 'play and rewards' then the 'win-lose' concept in games are considered. The procedure of play gives the setting for the probability assessments. These begin with 'starting the game' probability and a travel through to player and team single actions, then to repeats and exchanges, ultimately to the whole game and tournaments. Calculation can encompass all event conditions such as presence or absence of order and replacment issues, appropriate to particular games. As explained in Chapter 13 of Book One, for most problems two or more solution methods are presented. Some probabilities are given without working but they are elaborated on later. Simulation is employed at times and although using Excel functions for this is limited in respect of the number of trials (up to 1 million but often less) compared with the many millions in specific code (macros, programing languages) versions, it provides valuable information.

15.1.1 Playing games and probability

The ways in which games of chance and games that include a random element can be played have evolved in recent decades. The first original 'fundamental level' games are played with the physical apparatus and equipment, i.e. real dice, cards, roulette wheels, fruit (slot) machines, etc. Sports, such as tennis games and team play for football, baseball, or horseracing, are enacted on real courts with rackets, and real stadiums and tracks and of course, with real opponents and horses, etc.

Moving to another level, such games can be simulated in very simple ways, by replacing the participants with tokens or counters and the physical stadium, pitch or horse track with picture boards. Dice or another device provides the randomisation effect. There are many examples of such games and board games, found separately or in compendium versions, as well as many card games based on sport and other activities. At the home and recreational level, many of these games are played on a table (hence, the general name 'tabletop games', which we will use to differentiate from video and computer simulated games.

With sports and physical actions where the objective is to win or score higher points, skill is a major factor but a random element still influences outcomes. This random influence can be made more decisive by adjusting the ability of participants to a more equal footing, such as the handicap system in golf or putting more weight onto 'better' horses. The aim is to introduce more uncertainty - now outsiders have a chance of winning and the *betting* process for game and sport outcomes becomes less predictable.

With the advent of computers, all these forms of games, gambling and all the randomizing devices themselves can be simulated with much more facility. The element of chance may not be apparent to the user, but the games will be played according to these probabilities. A *random number generator* provides the randomness. No physical apparatus is required and there is unlimited scope within the basic concept of a particular game, e.g.

computerised versions of tabletop and board games, fruit machines, bingo, card games, sports, etc. on personal computers, tablets and smart phones.

15.1.2 Playing for points, tokens, money and betting

In many games, success can be gauged by the amount of winnings accrued during play or by that accumulated by termination of play. These 'winnings' need not be monetary in nature and are found as numbers or scores of points and tokens (false currency)' or as quantities of real money. Betting is usually associated with the latter form, but it can be applied to any situation where some amount is placed on the outcome of an event in the hope of gain. In game playing, there are two forms - one is where the player is a participant and actively places bets and sometimes makes the randomising actions during play, e.g. playing a card game with bets placed in the 'pot' for the winner.

The other is where the person making the bet is a spectator (the *punter* or *bettor*), has no involvement with the play and bets (usually with real money) on the result. This is true with all games, especially sports and there is also a long and enduring fascination in betting on 'the winner' of game contests, including those with no random element, e.g. chess tournaments.

Home and recreational play can be done for points and non-monetary units but vast commercial enterprises are based on gambling for real monies. The most well known of these are *casino games* and in sports, horse racing and baseball, soccer games etc. Prize levels vary depending on the setting and amounts wagered, ranging from carnivals and fetes that have smaller prizes, through casinos where winning can be as high or low as the gambler goes, to national lotteries that have mega–size jackpots.

Gambling, when played sensibly, can provide great entertainment and is enjoyed worldwide by millions of people. They enjoy not only the games, with their colourful lay out and equipment, but also the glamour, the luxurious trappings and surroundings of the casino or racetrack and the participation and interaction with other players. As is revealed in many texts (including this one) on gambling, all commercial forms have a negative return, i.e. overall, (unless you have big win), you lose more than you win. Many authors point this out but for some, playing the game has other rewards, even if a win has a 1 in 14 million chance. A game player can still get enjoyment from these games without gambling for real - many online versions (downloadable) of them allow play for free.

15.2 Winning with probability

Finding the 'best', or more appropriately, the **optimal strategy**, is the key to success in game playing. We do not cover game theory in this book, but at points, we will point out various suggestions for a probability-based strategy or at least some tactics to support strategic play based on probability and expected values. This advice can take the form of *'if you play the game this way you'll have a better chance of winning'* or it will illustrate and compare various styles in real or simulated play.

Turning to probability itself, there are various 'angles' for looking at 'win' success. We can consider individual player actions, player vs. player or banker / dealer and in sports, we have most of these plus interaction of teams. Single actions in isolation can be examined or we can look at a series of *repeated actions* within a section or at the game and match as a whole:

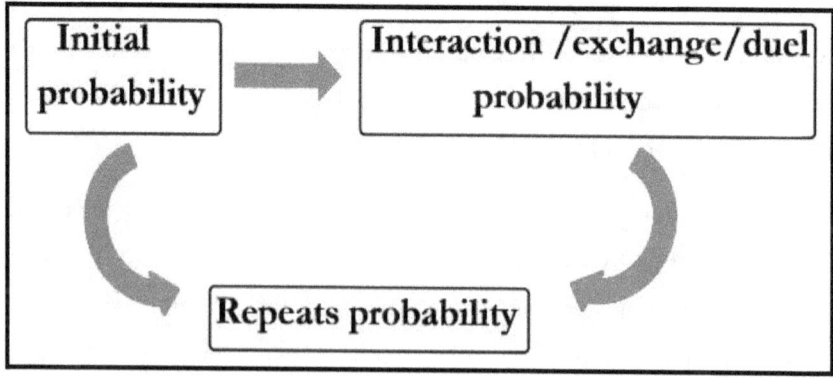

Additionally, we can look at the mechanics of play, such as *'who goes first or are actions simultaneous?'* Within all these, probabilities can be based on the randomizing device (classical probability) or for the action with the random element via counts and measurements, giving statistical estimates (relative frequency or others based on distributional models of past plays). Some of these aspects have been looked at in Book One and more follow here and in later chapters. As a first illustration, we take a look at the starting point for many games.

15.2.1 Starting procedures - 'who goes first?' or 'who starts?'

Many games or game stages require that one player (team) initiates the play and in the interests of fairness, this is often decided by a randomizing device. In sports, this is done by a coin toss, in board games by roll of 1 or 2d6, in other games with dice of other sizes and in card games by cutting the deck. This is a simple but important probability experiment, because 'who start's' may have a bearing on 'who finishes first' and wins (see below). It is a mini-game in itself with a loser and a winner (who initiates the play). It is also possible to decide this by employing a random influence, e.g. in darts *'nearest the bull goes first'*. If players are of roughly equal ability then the random element could be viewed as deciding who gets nearest on any particular throw.

Essentially, for all these problems, which are a form of duel, the ways of winning must be enumerated and compared to the total outcomes in the action to get the probability. With the coin toss, this is simple – each player or team has a 50% of winning and getting the option of choosing to go first. Throwing a die for starting a board game involves more work and can be solved in various ways.

Take two players (A and B) with 1 die (d6) - each player throws once and the <u>higher face value</u> wins and starts the game. *'What is the probability that A will win?'* Intuitively we might say 50:50 – is this true? Going through the process logically, it can be seen that A can win by getting a '6' with B getting a '5 or a 4, or 3 …or 1' = 5 ways. Likewise, A can win by getting a '5' with B getting '4, or 3 …or 1 '= 4 ways. Following these through and adding them all up (the ***addition principle***, Book One Section 7.1.1), we get 15 ways:

Spreadsheet Table 15.1

Dice throw events		
A	**B**	**A win**
6	1,2,3,4,5	5 ways
5	1-4	4 ways
…	…	3/2/1 ways

We are, in effect, summing a series of consecutive integers. Using a bit of maths, we can do this in a convenient calculation as,

Sum = X * (X+1)/2)),

where, X = the start and the series ends at '1' ***(15.0)***

Thus, A's win count = sum(5...1) = 5*(5+1)/2 = 15 ways (see also Fig. 18.5).

What we are doing is looking at the elements in the outcomes of 2d6. There are 36 of these and to make it easier to visualise, A could use a red die and B could use a green one. Looking at the count for A win events, we see that these outcomes (15) do not cover the whole sample space. The others cover 6 ways to get a 'tie' and another 15 ways where B wins, so the net result on a <u>single</u> throw each is:

$$P(\text{'A wins'}) = {}^{15}/_{36} = {}^{5}/_{12} = 0.416667 \ (42\%)$$

This procedure views the throws as *simultaneous* and takes them as a single throw of 2d6 (in the game *Risk*® ©Hasbro Inc., a similar situation arises where dice can be viewed as one overall throw). To confirm the above counts, readers can consult Table 4.5 (Book One) where outcomes for 2d6 are listed. There are 15 sequences where A beats B (where A takes the first element in each ordered pair), etc.

This is not 50:50 but if we consider the ***reduced sample space*** (Book One

Section 4.2.1*)* and ignore the ties, then P is calculated based on a reduced total outcome count of 30:

P(*'A wins single throw each, no ties'*) = $^{15}/_{30}$ = 0.5 (50%)

A tie (P = $^{6}/_{36)}$ gives 'no decision' and re-rolls must be enacted. These could go on to infinity, but this is a **geometric series** (Book One Section 10.2), which we can sum,

P(*'A wins'*)

= $^{15}/_{36}$ OR ($^{6}/_{36}$ * $^{15}/_{36}$) OR ($^{6}/_{36}$ * $^{6}/_{36}$ * $^{15}/_{36}$) OR ($^{6}/_{36}$ * $^{6}/_{36}$ * $^{6}/_{36}$ * $^{15}/_{36}$) ... ∞

The *first term* is identified as $^{15}/_{36}$ and the *common ratio* as $^{6}/_{36}$, then applying (10.1c),

P(*'A wins'*) = $^{15}/_{36}$ /(1- $^{6}/_{36}$) = $^{15}/_{36}$ * $^{36}/_{30}$ = $^{1}/_{2}$

Calculations with other sizes of dice follow the same steps, as with two players throwing a d10 for an initiative roll in a combat game or when dice of mixed sizes are rolled to decide any required issue, with players of differing ability. In these games, 'who acts first' may be critical. For cutting a card deck, calculations require more thought and are dealt with in Section 22.2.

More is given on this topic in the sections on **duels**, below. Similar, but more involved calculations can be devised for *any action* with one or more randomizing devices and two or more players to give probabilities for this game step.

15.2.2 Play probabilities

Once the initiating player has been decided, play can commence and the probability of the first and following game actions become the focus, whether this is based on an action with a randomizing device or play with a sport instrument, etc. Usually, simple events apply.

Base action probability - randomising devices

The probabilities for the various devices may appear simple to ascertain at first sight. They have well-established 'base' probabilities and the action is used over and over in the game, one of the commonest being dice rolls for movement in board games. Rather than aimlessly following random outcomes, the player can perceive relative probabilities, which can allow advantages with other procedures. Other games or stages can have more specific targets required to achieve objectives and probability calculations may get more complex e.g. *Craps* needs a '7 or 11' on 2d6 (P = 0.22) to win on one throw, a 'street bet' in roulette needs 1 of 3 numbers on the spin (P = 0.08). A card player in rummy has a 5% chance of converting a 2-card set to a playable meld, etc.

Additionally, play probability can be the same for each player or it can differ. In the previous roulette example, another player's initial play could be for an outcome with a higher probability, such as an even number (P = 0.49). In a role-play combat game, one player could require '14 or more' on 1d20 to score a hit, whereas another player or opponent could require a '9 or more'. In card games, the game commences with dealer doling out a random selection of cards. Each player has the same initial probability of getting any particular type of hand. The same perspective applies to player allocation or selections in other games like lotteries and bingo, etc. These and other random devices are detailed with fuller explanations in later chapters.

Base action probability - sport game actions (player or team)

For sporting actions, probabilities are not necessarily established or available and different procedures are required. This also applies to games involving physical skills, such as skittles, tiddly winks, marbles and shove ha'penny, as well reaction times in video games, etc.

Here, game probabilities are based ideally on recent data on the current 'form' of the player or team. The simplest assessment of ability is via a *statistical estimate* of past performance (as done in Book One Section 6.2,

Table 6.1). Data are accumulated as simple counts of success / failure of the action. The probability estimate is obtained by a relative frequency (rf.) calculation. That is, predicting future performance is estimated based on past actions, then modified according to current conditions, etc.

Alternatively, enhanced measures[1] can be taken for the same actions. With these detailed data, a distributional fit can be formulated, assuming approximation to the normal or other distribution (see examples in Book One Section 12.2). There are many variables and many models have been devised for sport circumstances. From these, probability for various actions in real and simulated games can be obtained: serving in racquet games, hitting on targets, scoring goals in team matches, ball sports and hockey, golf putting, getting over a certain distance in athletic and throwing sports, record times in descending a ski slope or getting round the hardest car race track in a motor sport console game, breaking the tiddly wink long distance on a single squidge, etc. Table 15.1 shows some examples where we can explore these facets, involving the actions of a representative casual and more serious-minded player or top team cf. a lower one on the league table for sport. Each carried out 100 actions, ranging from a single task, e.g. a throw, to a more drawn out series in a whole game, or in the case of the racing game, 'driving'.

The first entry in the table deals with the game of darts. The target is the centre of the 'triple 20' bed on the board. The target area is wider than it is deep but we assume that to score, the thrower needs to get within 5 mm. The relative frequency outcomes (hit or miss) were counted and gave probability estimates of 42% and 5% for the professional and casual player, respectively.

More in depth measurements were taken as the radial distance from the target (from the centre of treble 20 bed to where the dart landed). Assuming adherence to a normal distribution, the average and standard deviation were

---

[1] This refers to data recorded at a higher *level of measurement*, which contain more information

calculated for these data using Excel functions (AVERAGE( ) and STDEV.S( ), respectively) as illustrated in Book One Section 12.2. Here, the average value locates the region where the darts were centred, ranging from 0 (the centre of the target area) to approx. 170 mm at the edge of the scoring area. A hit is achieved by a dart striking within the 0 to 5 mm region. The standard deviation can be taken as a measure of skill mixed with random variation (in addition to any physical and psychological effects causing variation there is always a contribution of random error (Book One Section 3.6)).

Table 15.1 Estimates of probability for individual player or team game, actions based on 100 repeats

Action	Serious player / team				
	Count	rf.	ave.	SD.	P
Hit treble 20 at darts	42	42%	6 mm	3 mm	37%
Team win / win by point lead	team A 63	63%	2	1	98%
Long distance squidge (> 5m) in tiddly winks	78	78%	5.1 m	0.26 m	65%
Complete top level track in console car racing game	38	38%	152 sec.	7 sec.	36%
	Casual player / team				
Action	Count	rf.	ave.	SD.	P
Hit treble 20 at darts	5	5%	29 mm	14 mm	4%
Team win / win by point lead	team B 49	49%	1.5	3	69%
Long distance squidge (> 5m) in tiddly winks	14	14%	4.3 m	0.56 m	9%
Complete top level track in console car racing game	20	20%	159 sec.	13 sec.	24%

* Note that the models are simple

We can see that the professional's dart distribution spread is 'tight' with a very small deviation (3mm) centred at 6mm and almost half the shots hit the treble. The casual player is off the target by a larger distance (29 mm), has much more variability (deviation of 14 mm) and consequently only five darts hit the target. The NORM.DIST( ) function was used to get a probability for a successful action, giving 37% for the professional and 4% for the casual player, a slight refinement of the relative frequency estimates,

P(*'up to 5mm from target'*) = NORM.DIST(5,6,3,1) = 0.369441

P(*'up to 5mm from target'*) = NORM.DIST(5,29,14,1) = 0.043238

For an example that demonstrates more effectively the superiority of the more detailed data, we return to the darts example referred to in Chapter 6 (Book One): two dart throwers, one (player 1) with a 50% rf. estimate of hitting the bull and another (player 2) with 30% but less scatter. These were compared on the radial distance measure described above, giving 25% and 31% probabilities, respectively. This reverses the order of ability and illustrates the greater discrimination of the distance measure, in respect that player 2 was more accurate.

Similar treatment can be given to location of tennis serves, cricket bowling, etc. to give data that are more accurate on which to base calculation for actual game and match outcomes.

Previously, we have used a win:lose relative frequency comparison for a team game (Table 6.1, Book One). This could apply to any team game such as football, ice hockey, baseball, etc. With such games where goals, runs or points are scored, valuable data lie with totals, averages, point leads and many other measures of more specific detail, such as shots from a distance, number of passes to convert to a score, etc.

Take the example of team play in Table 15.1, based on relative frequency of wins over a season (assume 100 games) with other contenders in the league. Team A has a rf. of 63 % of winning and team B 49%. The scores were recorded, so it was possible to take the average and standard deviation of the *margin of win (difference* between the team's score and its opponent's*)*. The latter form of data can be used to model according to the normal distribution but also others such as the Poisson distribution, viewing the goals as isolated random events in a long time period (Stern). This measure can range from negative, as in the case of losing, through zero (a draw), to various degrees of positive depending on the final scores. The relative scores for A averaged out (over all the opponents in the league) at a small positive value (ave. 2, SD 1), whereas B had a lower margin 1.5 and more variability (ave. 1.5, SD 3). Using the NORM.DIST( ) function, the probability of achieving a win is represented by an average of > 0,

P(*'Team score point lead +1 or more'*)

Team A = 1- NORM.DIST(1-1,2,1,1) = 0. 97725

Team B = 1- NORM.DIST(1-1,1.5,3,1) = 0.691462

The analysis shows that team A is almost certain (98%) to get above 0 (i.e. win) and the majority of games are >2, but with team B – getting >0 is less probable (69%). Measures like these can be obtained for all teams in the league and relative standing of 'win ability' used for further calculation of possible championship candidates.

This model serves to illustrate the principle, but it is limited when compared with others, which include adjustments for home / away games and the team's defence and attack ability, etc.

With batting sports, such as baseball or cricket, relative frequency can be used to count hits that score (any type of run) and those of a particular type such as a home-run or a 'six', respectively. A more comprehensive 'batting average' can also be calculated, the method depending on the sport (Winston, en.wikipedia.org). Based on these, players can be classified as relatively high or low 'hitters' and comparison made with win rates for the teams etc.

With distance measures, data can be gathered in their simplest form as a count of actions reaching to and over a designated distance in the same manner to that for javelin throwing (Book One Section 12.2). As a complete contrast, two players in the recreational game *tiddly winks* (which has a long distance record; www.etwa.org/records.html) were compared. One, more dedicated, squidged 78 winks 5 meters or more out of 100, the other, more of a fun player, managed only 14 over this distance. Thus, rf. estimates are 78% and 14%, respectively. The distance attained by each wink was measured and analysed to get the average and standard deviation of the data. Assuming adherence to a normal distribution the function gives,

P(*'p1, squidge over 5m'*) = 1-NORM.DIST(5,5.09,0.26,1) = 0.649634 (65%)

P(*'p2, squidge over 5m'*) = 1-NORM.DIST(5,4.25,0.56,1) = 0.0916720 (9%)

The keen player has a 65% chance of getting over 5.5m, with the other player trailing well behind at 9%. Both of the normal distribution probabilities are lower than their rf. equivalent.

For the video game console, the dedicated player managed 38 out of 100 completions of the top-level racetrack (rf. of 38%) less than the qualifying time of 150 sec. Average lap time of 152. sec. (SD = 7). Using the time data showed that successful laps were less than 150 sec. and that competitors were clocking in 150 sec. at their maximum. The NORM.DIST( ) function can give the probability of each player getting 150 or less, as,

P(*'at most 150 sec.'*)

p1 = NORM.DIST(150,152,7,1) = 0.362890 (36%)

p2 = NORM.DIST(150,159,13,1) = 0.240586 (24%)

The dedicated player has the necessary speed, but his friend is slightly slower with fewer laps below 150 (20) and 159 sec. on average and more variable (SD = 13). Both players are in the poor to moderate region for getting below 150sec. Player 1 with more dedication could achieve a greater than 50% chance if he can reduce average lap time by 3 sec.

Many of the above established initial play probabilities are single actions. They may, on their own, be the main deciding factor in 'who wins', unless equal for all players and whoever has the first move may have an advantage (see below). Otherwise, they may follow through in the game and make up the probability basis in repeat actions and exchanges.

15.2.3 Interaction probability - repeats, exchanges and duels

The game has started and depending on its nature, it can end rapidly or last for many hours. Once the above determinations have established probability for the action of interest, calculations for carrying out the action repeatedly, or exchanging actions with opponents, can proceed. Possible modes of play are:

- One player repeating once or more, i.e. pl then pl again then pl again
- Exchange in sequence - player 1 then p2 then pl p2 (in order)
- Exchange as one (simultaneous)

These give rise to combined actions in the form of *sequence, content* and *run events* (Book One Section 5.3). This process continues until an end point in the form of an objective or score is achieved and in some games 'who gets there first' wins. Probabilities for combined consecutive actions (*sequence events*) are usually the simple product of the base probabilities assuming each action is independent (*the multiplication rule* (5.1a) ('AND'), Book One Section 5.2.1). Order can be important, as in the alternate action case. If successful actions occur within a longer series in any order, interspersed with fails and successes of the other player or team, then the *binomial* and *negative binomial* functions (Book One Sections 9.2 & 10.3, respectively) can be used for *content events*.

For randomising devices, many games have sequential turns but sometimes a player may gain a temporary repeat action, where success allows continuation of the turn. That is, the player can win a single action and proceed to a following stage, continuing progress to a further point or an outright win, without involving the other player.

For example, in board games, certain dice outcomes allow an extra turn ('6s' or 'doubles') or when an opponent is penalised. In the dice game *Pig* (Section 20.3), a successful player can go right through to win on one turn (P = $^5/_6$ * $^5/_6$ * $^5/_6$ ...). In some card games, a player has the option of continuing play on a turn, based on previous success, e.g. in *Higher or Lower* (Section 22.2), a successful choice allows the player to let the bet ride for one or more turns (in this case probabilities can change but this can be viewed as a series of successes). In *Roulette*, a player can keep the same bet running, e.g. P(*'Four blacks in a row'*) = 0.49 * 0.49 * 0.49 * 0.49 = 6% (note: this is different from getting 'four black out of 9 spins', P = BINOM.DIST(4,9,0.49,0) = 25%).

Many cases of this are found in sports and other games. Take a *pool* ball potting contest. Player 1 starts and pots 1 ball then a 2nd and goes on to the last - player 2 does not get a turn. If player 1 has an average estimate of 95% potting ability, then such a player could pot 10 balls in succession, P = 0.95^{10} = 60%. Or, a tennis player can win a whole game on 4 successive serves if each one is an 'ace'. A player with a chance of 12% of an ace could win a game on first serve, P = 0.12^4 = 0.000207360 (not much chance but it does happen). This assumes that each serve is independent and left and right side ace serves have the same probability.

Take the professional darts player above. We have established a probability of 37% to hit the treble 20 bed. The get all three darts in for the famous '180!' score is much more difficult and has probability,

P(*'treble 20 in 3'*) = 0.37 * 0.37 * 0.37 ≈5%

Thus, once in every 20 visits to the oche will get a '180' (here the independence assumption may lessen as the target area gets smaller). The casual player has much less chance ≈ 0.01% (once in every 10,000 attempts.)

Of course, fails can be repeated as well and these calculations apply not just to an action but also to the whole game or match (see below).

Repeated actions as streaks for player or team

The ultimate illustration of repeat actions is when a player or team has a *streak* (or *run*) of successes or failures. This can apply within a game or to many games over a period of time. From sports to 'snakes and ladders', winning several times in succession is taken as a sign of good luck or luck combined with skill (lack of), respectively, depending on the game. A corresponding string of failures has opposite connotations. However, caution is advised with this viewpoint, as streaks can occur purely by chance.

Examples of this event form were introduced in Book One Section 5.3.3. Some of the examples above are streaks in the 'isolated run' version (obtained by p^n), viz. 'four blacks in a row at roulette (= (BBBB)) or 10 pots in a row at pool. A gambler who gets such a run of blacks in roulette, may think that he or she is on a winning streak, but if you examine a list of previous plays (say 100), a streak of four or more usually occurs at least once. Streaks of successes or failures occur in team matches. Take another team (C) in the league above that has a 30% win chance based on their average margin. Matches have started and C has lost 5 games in succession, $P = 0.7^5 = 17\%$, not high but certainly possible, resulting in a swift move to the bottom of the league table. Is it just bad luck, or is C a poor team? Looking at the games as a binomial experiment (n=5, p = 0.7), C would expect to lose 5*0.7 = 3 to 4 games, with SD ± 1, so it doesn't look high and C is performing 'on par'.

The same calculation approach is used for randomising devices for many examples of the streak or run in its various versions. Illustrations for coins, dice, cards etc. appear later. Rather than for devices, wins or loses for the whole game may occur in streaks, as with the team C example above. Other examples of 'whole game' streaks are: being dealt a blackjack on 3 consecutive deals $P = 0.48^3$, or losing your favourite board game, where you 'usually' (75%) win, 5 times in succession $P = 0.25^5$.

The streak can occur mixed in with failed actions and other streaks, with consequent increased complexity for the probability calculation. A simple mixed run (formula (5.3)) has one streak along with fails and examples appear in later chapters. However, with a long sequence length and a short streak, a simple run would have a very low probability and it is much more likely that there would be mixture of fails and successes. Additionally, we can see that there is plenty of opportunity for other streaks of equal or greater length to start. While it is possible to extend formula (5.3) for a more complex mixed run for some examples (see Problems & Exercises), with longer sequences, we need mathematics that is more complex or we can take approximations. The Poisson function (Book One Section 9.3) can be used as an approximation for these probabilities, where there is a large number of trials and the isolated streak probability is low (Gould). This assumes the streak is a rare event over a long time period or series of actions.

To illustrate the method, take Wally the demon cricket bowler, who, during practice, bowls at an undefended wicket and previous data show an estimate (rf.) of a hit on the stumps at 0.4 (40%). *'What is the probability estimate for six 'bowled outs' in a row during 100 bowls?'*

The approach is to calculate the probability of the isolated run, then view it as an event in the sequence or content, reduced by the streak length. For the isolated streak (length =6), the probability is given by,

$$P = p^n = 0.4^6 = 0.00409600$$

The next stage is to calculate how many opportunities the streak event has to happen. Although there are to be 100 actions, not all of these will allow a streak of length 6 to start and end, in particular the last 5. So, the number of 'slots' for the streak to fit in starts at the bowler's first bowl and continues to the 94th. The total number is given by,

$$100-6+1 = 95$$

Thus, in 100 bowls, this streak could fit in at 95 positions in the series. However, these opportunities must, apart from the first, start with a failure (P = 0.6), so the number we would expect would be $= 1 + (n-r) *q = 1+ 94*0.6$

= 57.4 (if p is very low, then q will be high and this step has minimal effect on the approximation).

Before we can use the Poisson function, we need the average number of streaks, and for the example above,

lambda (λ) = 57.4 * 0.00409600 = 0.2351104

Therefore, to get 1 streak of 6 successful bowls mixed in with fails and other bowls,

P('*1 or more streaks of length 6 in 100 bowls*')

= 1- POISSON(1-1, 0.2351104,1) = 0.209516 (21%)

Consequently, the bowler has a poor to moderate chance of at least one of such a streak in his 100 bowl session. The same procedure can be used with other sport actions such as free shots in basketball or netball, goal shots in hockey, football sports, etc. The overall probability of the occurrence of streaks in sports can also be looked at over many games and seasons for players and teams or as an event within the historical statistical data of the sport with many thousands of trials. Thus, although some long streaks have very low probabilities, they do occur (Winston).

With randomising devices and streaks, the number of trials may be too low, but long sessions can occur in some games - how many deals are there in a long card playing session? Say a marathon poker session has usually 250 deals. You know that the probability of a 'straight' is low (0.00394), so what's the probability of getting dealt 2 straights in row?

P('*isolated streak*') = 0.00394^2 = 0.0000155236

No. opportunities = 1+ (250-2) * (1-0.00394) = 248.023

lambda (λ) = 248.023 * 0.0000155236 = 0.003850208

P('*One occurrence*') = 0.0038428054

It is extremely rare to get one straight after other. However, there are hands with higher initial probabilities. The same calculation for a pair (P = 0.422) gives virtually zero probability for exactly 1 streak (of length 2), but 99.9999... % for 1 or more, i.e. you are virtually certain to get dealt a pair twice in a row, one or more times, over 250 deals.

In games where dice are part of the game equipment, runs of face values will occur. The more rolls in a typical game, the more likely the run depending on the game. In *Ludo*, there is a good chance of a small run of '6s' (Section 20.2). In large board games and long lasting campaign adventure and role-play games, dice rolls may accumulate in number. In combat battles and melee, getting streaks of certain outcomes can be critically important for success or failure. With misfortune occurring in runs as well, even short runs of certain outcomes can ensure victory or defeat depending on the roll (see examples in Sections 18.3.4, 20.2 & Problems & Exercises).

Two or more player exchanges - duels & 'who gets there 1st?'

Once exchanges begin, then both (all) player action probabilities and estimates must be taken into account. We assume that one player does not dominate as described above where some single actions can decide a point and a series of these can occur without input by the other player. In games where there is order of play, there will be at least one action by one player before the end but more can occur in a series of *exchanges.*

This interaction of plays, turns, goes, shots, etc. can be viewed in the simplest forms as *duels* (truels for three players, etc.). The same probability structure applies for any duel exchange – whether it is tossing coins or shooting at targets, etc. Play actions can be *simultaneous* as in 'firing at the same time'. These can be 'sudden death' games with the action(s) comprising the whole game or they can be a prelude to further play. Exchanges can result in nothing happening and a second round or turn follows, etc. until the end is reached. This is in the form of a target pattern, as with *'first to get two Hs'*, designated scores *'first to get 501'* in darts, first to get to 11 points in squash.

So, the greater total, points, stages completed, touch- downs, goals, baskets, hits, etc. signify termination of play. The stopping point concludes the particular stage or the whole game and whoever performs this final action wins (or a draw is declared). We will look at some aspects of these probability structures limiting the interaction to two player scenarios.

Duels - exchanges in sequence or together

Duel probability illustrations typically take the case of protagonists exchanging shots. Consider two paint ball or laser gun shooters who have probabilities of hitting the target or each other of 0.4 (A) and 0.3 (B) respectively. What chance has A of winning on <u>alternate</u> shots?

The problem is characterised by being possibly infinite in nature – i.e. if both shooters miss they will continue to fire at one another for an infinitely long time, given unlimited time and ammunition, etc. As pointed out above 'who fires first' is critical as this player has first chance at a win outright before the opponent gets a turn. This first scenario assumes A goes first and a hit ends the exchange:

Spreadsheet Table 15.2

1st exchange events				
Outcome		Result	Calculation	P
Event X	A hits		A wins	-	0.4
Event Y	A misses	B hits	B wins	0.6 * 0.3	0.18
Event Z	A misses	B misses	No-one wins	0.6 * 0.7	<u>0.42</u>
					1

On the first exchange or round, we have one single outcome and two <u>ordered</u> pairs (probabilities for the latter are calculated according to the *multiplication rule*). One of the above events must happen, and the most likely one is that they both miss followed by a hit, but A has a better chance than B. For an A overall win, we apply *the addition rule* ('OR', Book One Section 5.2.2) and add the possibilities that result in a hit for A:

Spreadsheet Table 15.3

	Outcome				Result	Calculation	P
E1	A hit				A wins	-	0.4
E2	A miss	B miss	A hit			A wins	(0.6*0.7)*0.4	0.2
E3	A miss	B miss	A miss	B miss	A hit	A wins	(0.6*0.7)*(0.6*0.7)*0.4	0.07
E4	...	...	...	...	...	...	...	...∞

As seen, this is an infinite series and we can use the method described above for a **geometric series** to get the summed probability. The *first term* is 0.4 and the *common ratio* is readily identified as (0.6*0.7). Thus using (10.1c),

$$P('A\ 1^{st},\ A\ wins') = 0.4\ /(1-(0.6*0.7) = 0.4/(1-0.42) = 0.689655\ (69\%)$$

Player A has a good chance of success in this encounter, but what about B? (see below).

The expression for the player who takes the first action can be generalised with the base probabilities for the opponents as 'a' and 'b', respectively, then with a bit of algebra,

$$\textbf{\textit{P(A 1}}^{st}\textbf{\textit{, A win)}} = a/(1-(1-a)*(1-b)$$
$$= a/(1-(1-b-a+a*b) = a/(0+b+a-a*b)$$
$$\textbf{\textit{= a/(b + a - a*b)}} \qquad\qquad \textbf{\textit{(15.1a)}}$$

Similar procedures can be used to get expressions for a 'B win' under the same or different conditions and with simultaneous actions, etc. One of the calculations appears as an exercise (Problems & Exercises) and from all these we get:

For a *'B 1st B win'* circumstance, (15.1a) applies, i.e. B becomes A, but to avoid confusion,

$$\textbf{\textit{P (B 1}}^{st}\textbf{\textit{, B win) = b/(b+a-a*b)}} \qquad \textbf{\textit{(15.1b)}}$$

With the above data (a = 0.4, b = 0.3),

$$P = 0.3/ (0.7- 0.4*0.3) = 0.517241\ (52\%)$$

Thus B's chances, even firing first, are markedly less than A's, due to a lower initial probability.

B's chances are even lower when she remains the second opponent,

$$P(A\ 1^{st},\ B\ win) = b*(1-a)/(a+b-a*b) \qquad (15.1c)$$
$$= 0.3*(0.6)/(0.7-0.4*0.3) = 0.310345\ (31\%)$$

Thus, B's place in the firing order is not enviable as her success probability is roughly one third of A's. In a 'fast draw' situation, possibly B could gain some ground (see in Problems & Exercises). Note that these figures sum to unity, as if both miss they continue firing, i.e. the formulae consider continuation until the end-point and do not give 'single exchange' probabilities.

Applying this to other games is reasonably simple, e.g., two players throw dice –the winner is *'first to get a '6'*. A starts and a and b $= \frac{1}{6}$,

$$P('A\ 1st,\ A\ wins, a=b') = a/(b+a-a*b)$$
$$= \frac{1}{6}/(\frac{1}{6}+\frac{1}{6}-\frac{1}{6}*\frac{1}{6}) = \frac{1}{6}/(\frac{2}{6}-\frac{1}{36}) = \frac{1}{6}/\frac{11}{36} = \frac{1}{6}*\frac{36}{11} = \frac{6}{11}$$

The probability for B to win, if she rolls first, would be the same, but if second,

$$P = \frac{1}{6}*\frac{5}{6}/(\frac{2}{6}-\frac{1}{36}) = \frac{5}{36}/\frac{11}{36} = \frac{5}{11}$$

The sample space is unevenly divided by A's and Bs' probabilities because of A's advantage.

The formulae can be simplified for a = b and,

$$P(A\ 1^{st},\ A\ win,\ a=b) = 1/(2-a)$$

and

$$P(A\ 1^{st},\ B\ win,\ a=b) = (1-b)/(2-b) \qquad (15.1d)$$

Thus, with the above shooting example, except that a = b = 0.4 $(\frac{4}{10})$, then A still has the advantage,

$P('A\ 1^{st},\ A\ wins') = 1/(2 - {}^4/_{10}) = {}^{10}/_{16}$ cf.

$P\ ('A\ 1^{st},\ B\ wins') = {}^6/_{10}/\ {}^{16}/_{10} = {}^6/_{10} * {}^{10}/_{16} = {}^6/_{16}$

With a coin game stage, first to get 'H' starts, a=b= $^1/_2$ and,

$P('A\ 1^{st},\ A\ wins') = {}^2/_3$ and

$P\ ('A\ 1^{st},\ B\ wins') = {}^1/_2 * {}^2/_3 = {}^2/_6 = {}^1/_3$

It seems that if the game's nature allows, going first gives an advantage.

This applies in *without replacement* conditions in games, e.g. drawing alternate cards and first to get to a particular rank or suit wins. The calculations are similar but the method is based on a situation comparable to the geometric but *without replacement* (Section 22.3).

Duels - simultaneous actions

If the actions occur together they are viewed as *simultaneous*, and *ties* enter the equation. This also applies when the game rules specify an inequality between two or more outcomes (see below).

For the simultaneous exchange, we assume that failure by both does not halt the game but any hit(s) does. The first exchange possibilities are viewed as <u>combinations</u>, i.e. no order of first and second actions, each comprising the two actions. This gives,

P(Sim., A wins) = a*(1-b)/(b + a - a*b)   *(15.2a)*

This expression gives the probability for any number of turns or rounds, continuing as long as no hits occur. With the data above (a = 0.4, b = 0.3),

$P('Sim,\ A\ wins') = 0.4 * (0.7) / (0.4 + 0.3 - 0.4*0.3)$

$= 0.28/(0.7-0.12) = 0.48276$ (48% ;cf. 69% with A 1st)

For the probability of B winning, 'a' switches with 'b',

P(Sim., B wins) = b*(1-a)/(a + b - b*a) *(15.2b)*

$= 0.3 * 0.6 / (0.7 - 0.3*0.4) = 0.31034$ (31%)

Consequently, B's lower initial probability means that A still has the edge when both fire as one and although it may not feel like it to B, the formula is equivalent to (15.1c) where A fires first. This time, the two probabilities do not add to unity - why is this? (see below).

If a = b, the simultaneous action expression becomes,

$$\textbf{\textit{P(Sim., A win, a=b) = (1-a)/(2-a)}} \qquad \textbf{\textit{(15.2c)}}$$

Under these conditions,

$$P('Sim., A \text{ wins, } a=b=0.4') = 0.6/(1.6) = 0.375 \ (38\%)$$

This time, A (and B) have a 38% chance. With the two player dice throw to see who gets '6' (or '1') first to continue a game, $a=b= \frac{1}{6}$,

$$P('Sim. A \text{ wins, } a=b \ \tfrac{1}{6}') = (1-a)/(2-a)$$
$$= \tfrac{5}{6} * \tfrac{6}{11} = \tfrac{5}{11} \ (\text{cf. } \tfrac{6}{11} \text{ when A goes } 1^{st})$$

The same probability applies to B's chance. Note the effect of combinations – the sample space is $\frac{11}{11}$ and there is an equal chance ($\frac{5}{11}$) for each player to win. Additionally, we can see that with this example and the prior ones, there is a possibility of both hitting (a tie or draw, P = 21%, 25% and $\frac{1}{11}$, respectively). For the dice example, this can occur in one way: each player rolls a '6' (see in more games and pre-actions in later chapters and below). There are other outcomes where nothing happens (they both miss) but the formula assumes continuation until one or more hits.

For an *inequality* in the simultaneous firing case, we cannot assign 'a' and' b' probabilities as the chance of win depends on a contribution of a random action from all (each) opponents. Results are decided by comparison of outcomes, i.e. participants do not have a specific target outcome and it is the one that qualifies in respect of being 'higher' / lower/ greater, etc., that governs the exchange. Decisions cannot be made by one player's action - both (all) must make an action - one player cutting the card deck is not enough to decide a bet or game start - both must act. This occurs in various other games particularly with dice - *Risk®*, *Dice Wars (Games Design)* and other war games where both players roll and whoever gets the inequality, wins. Usually, one player is called the attacker, the other the defender - in a turn based system. Additionally, the action can be anything from a 1 object / 1 action state to many objects in a compound event, e.g. in *Dice Wars,* 8d6 vs.7d6, or in a campaign war game, 30d6 vs. 26d6. Ties can occur, but some games avoid this by deeming a tie as a fail for the attacker.

The calculations for the probabilities of a win are more complex and lengthy for these situations. Return to the case at the start of this chapter, where two players throw dice to see who starts a game (Section 15.2.1). Throws are viewed as simultaneous, but an inequality is stated ('whoever is higher goes first') and the duel formula does not apply and the probability needs to be calculated via the infinite series as described above.

The above formulae do not cover all outcomes for duels, because of course, other things can happen, such as both can hit and both can miss or the duel could end at a particular point, etc. (see Section 20.2.2 and Problems & Exercises).

The complete game, matches and tournaments

In some cases, player probabilities and other measure effects can be built up to give the *win probability* for the game, for the player or team, as done with some examples above. Many games have turns or phases, reflecting passage of time or exploration stages or objectives. This can also be seen easily for sport games that are enacted in stages or rounds, e.g. tennis with its set, game and match structure, frames in pool and snooker, innings in ball games, etc. All can have their own probability and be linked together to give overall win probability. This can be in sequence or as content via the binomial function.

In tennis, player A when serving has probability of 45% of winning the point, so B by defending has p = 55%. If we build these values up into the game structure, we get an expression for A to win a set as,

$$P = p^4 \ \text{OR} \ q*p^4 \ \text{OR} \ q^2*p^4 \ ..., \text{etc.}$$

The same process is applied to the following sets, ultimately to the game then to the whole match. The calculations get increasing complex (Epstein, Haigh) but the principle of the method is clear.

Team A above, has a win probability (based on 'margin of win') that calculates out at 98%. To win all 8 games in a series, A has $P = 0.98^8 = 0.850763$ (85%), but team B at 69% has poor chance 0.0513798 (5%). Team A looks like a strong contender for league champion.

The whole game can be viewed in this way. Jack has a good record (67%) at winning in 2-player canasta - he won the last 3 weekly canasta evenings against Jill ($P = 0.67^3 = 0.300763$ (30%).

Bert seems unlucky at bowling. His strike incidence on previous sessions shows he gets a strike 6 times out of 10. This session he has bowled 20 times and got 4. Based on his strike estimate getting only four would be BINOM.DIST(4,20,0.6,0) = 0.000269686. On 20 bowls he would expect on average, 20*0.6 = 12 strikes and he has a 60% chance of getting at least 11. Bert must be off-form.

Alternatively, a macroscopic view can be taken based on the relative performance for one player or team over others. With team games, performance assessment is according to past wins, draws and loses, which leads us to probability in betting.

15.2.4 The spectator punter / bettor

Imagine Bill the betting fad, who, likes to have a wager on team matches and an occasional flutter on racehorses. Probability and the odds expression were discussed briefly in Book One Section 2.2.2. We will look at this in the current section and readers are referred to the Probability Calculator in Spreadsheet Table 3.1 Book One Section 3.5, which is useful for quick conversion of decimal probabilities to odds.

Team matches

For team games, Bill is aware that the system used by bookmakers in assessment is according to past wins, draws and loses. He supports team A ('Alwayslose Rovers') and next Saturday they are playing a home game against B ('Bagagoal United'). He looks up his records of previous home games (11) and finds that his team has not been doing well - they have lost 5 of these and drawn 3. The team they are imminently facing, 'Bagagoal', has done better playing away:

RESULT	A	B
Home wins	3	-
Away wins	-	8
Drawn home	3	-
Drawn away	-	2
Lost home	5	0
Lost away	-	1

Simplistically, Bill could use a win% to gauge the chance as P('A home win') = $^3/_{11}$ of a win for A, (P('B away win') = $^8/_{11}$), and correspondingly for a 'home draw' $^3/_{11}$ and 'home lose' $^5/_{11}$, but this ignores the opposing team. A more informative method (www.goal.com) similar to those of the bookies, uses all the data above:

A 'win at home' = A home wins + B away loses = 3 + 1 = $^4/_{22}$ = $^2/_{11}$

'Draw' = A home draws + B away draws = 3 + 2 = $^5/_{22}$

B 'win away' = B away wins + A home loses = 8 + 5 = $^{13}/_{22}$

These totals are all divided by the total number of matches (22) to get probabilities as fractions.

These odds give a slightly different picture to the simple ones above -they are less for win and draw, with more weight added to a B win. The method used is still relatively simple compared with others that use point allocation for attack, defence and other factors and models based on probability distributions for deciding team game winners.

Are the odds above complete? Well, the procedure above encompasses some aspects of the two teams' performance, but there are many more. The bookmakers, and Bill if he has the information, may want to adjust these odds based on relevant up-to-date material, concerning the teams (players, manager), weather, etc. Bookies also build in their percentage (*'over-round'*) and they may look at other books and general betting before they finalise. Let's say the bookmakers produce initial odds based on how they are advised by their expert consultants, etc. Their odds (fractional) are similar but slightly lower - they rate Bill's team at '10 to 1 against' ($^1/_{11}$) for the match:

	Bill's odds	Bookies odds	P	Odds against	Over-round	Final odds
A win	4/22	2/22	0.090909	10 to 1	0.100000	9/1
Draw	5/22	6/22	0.272727	8 to 3	0.300000	7/3
B win	13/22	14/22	0.636364	0.6 to 1 (3 to 5)	0.700000	3/7
			1		1.1	

The first part of the table would be a *fair book* and if offered, the bookmaker would not make a profit. *Over-round* is applied by rescaling these bets to cover more than 100% - in the example to 110%. Odds are recalculated and it can be seen that they have been reduced. No matter what the outcome, a small percentage goes to the bookmaker (final odds are expressed in UK bookmaker form (these can be calculated in various ways. Perhaps the simplest is to express the winning probability as a fraction then subtract the numerator from the denominator to get the odds against (losing) and express as losing/winning, simplifying the fraction if necessary but retaining whole numbers on both sides, e.g. P = 0.3 = 3/10, gives (10-3)/3, '7 to 3' or = 7/3).

Bill decides to take this bet as he feels that this team has a better chance than the final odds of 9 to 1. On winning, he would receive 9 times his stake plus his bet. If he deemed that this was too much of a risk and went for the draw, he would get £7 for every £3 bet plus stake. His friend Alphonse feels that Bill is just a bit too loyal and bets £10 on a win for B - winnings are the other way round and he would receive £3 for every £7 wagered along with his bet (£4.29p + £10). These are *fixed odds* and remain as they are when offered. As the season progresses, the odds are recalculated according to performance.

Of course, bets are available on other aspects of team games, such as goals, shots, points scored, point leads, etc. and 'who is going to win the championship?' For this, the standing of all teams' performance is taken into account to give an overall ranking and fixing of odds. These may change as the season progresses, but bets placed at start are honoured. In this way Bill may get very good odds for a championship win for his team, say 100-1. Even if the Rovers begin to perform better, indicating an improved chance, his original bet is guaranteed. Who knows perhaps they will pull off some shock results!

Probabilities for other types of games with tournaments and elimination rounds or stages can be calculated in a similar manner or just based on current form win%. In the local bridge championships, partnership A has won 60% of their games. Four rounds remain to be played. Assuming comparable circumstances,

$$P('Partnership\ A\ wins\ championship') = 0.6^4$$
$$= 0.129600\ (13\%),\ odds\ of\ 7\ to\ 1\ against$$

This type of calculation is common with sport tournaments. For example, in snooker with a series of 7 frames, the first to 4 wins. Player A has $p = 0.7$ and $B = 0.3$ for a frame win probability - what is A's chance of winning?

For a player to win they can achieve this by having won 4 frames by 4th, 5th, 6th or the 7th.

The number of wins (successes) is fixed and it is the number of fails that change, a **negative binomial** case, viz. A needs to win outright in 4, OR in 5 having the final win on the 5th, etc.:

P('*A to win*') =

= NEGBINOM.DIST(0,4,0.7,0)

+ NEGBINOM.DIST(1,4,0.7,0)

+ NEGBINOM.DIST(2,4,0.7,0)

+ NEGBINOM.DIST(3,4,0.7,0) = 0.873964 (87%) 7 to 1 in favour

As imagined, player B has a much lower chance, P = 0.126036 (13%).

Sometimes scrutiny of opponent ability can yield an advantage when play is in certain order and there is a choice of opponent order (Mosteller[a]).

Horse racing

Horse racing betting is a very popular form of gambling. The horse and the jockey are the ones engaging in the action, honed by training and experience. Horse racing simulation is popular in games, as with the board game *Totopoly*® ©John Waddington, Ltd, Hasbro Inc. As far as probability goes, and the game player, whether on the real live race course on in simulation, the interest lies in predicting which horse wins or is placed, although the board game has participation in training.

Bill is looking at an up and coming handicap race, where there are seven horses. On a crude basis, one would say that if we have seven horses and jockeys all of the same ability then there would be a $1/7$ (6 to 1 against) chance for any particular horse to win. Bill realises that this is unlikely and there will be variation according to many factors. The form and condition of the jockey and the horse in particular are paramount (Donato), win %, speed over last 4 races, average amount of money won per race in the year plus the venue and condition of the ground, etc. A basic win% is a starting point based on the last 4 races. Bill examines the data and they show that the horse

'Crosshair' has won 2 out 4 races and outranks all the others - so is this the favourite? Does everybody know this and will they all bet on Crosshair? Well, one point is that horses with a history of success could upset the betting process. It would become superfluous to a degree and punters would always bet on the better ones. However, Bill knows that this is overcome by a form handicap limitation, by adding weight to more successful horses.

Ok, but these horses still seem to win, even with extra weight, so other factors are contributing and Bill, as a budding 'handicapper', has work to do. This information may be available in part or whole to the punter, who needs to 'study form'. Thus, a horse that has been seen to be clocking fast times during training, may be noted as being on form and could be become an 'odds-on favourite', as could horses trained in stables with a reputation for winners. The odds above of '6 to 1 against' for all would be modified according to the weighting of the various factors gathered on the recent condition of each horse. Bill assigns points for each factor plus any subjective hunches and comes out with what he thinks is a more realistic picture (1- 100% overall for each horse):

	Horse	Bill's pts. %	Bill's odds	Bets placed	Tote odds	
1	Tabletop	10	9 to1	15	0.0789470	12 to 1
2	Fumble	15	6 to 1	10	0.0526320	18 to 1
3	Crosshair	35	2 to 1	50	0.2631580	3 to 1
4	Duboticus	20	4 to 1	25	0.131579	7 to 1
5	Spintrue	20	4 to 1	25	0.131579	7 to 1
6	Acehigh	5	19 to 1	30	0.157895	5 to 1
7	Last Oasis	10	9 to1	35	0.184211	4 to 1
				190		

So, are these odds comparable with those of the bookies? It is possible, but unlike the fixed odds for the team game above, a different system altogether is used.

The totaliser system

The *tote system* (en.wikipedia.org) involves calculation of odds based on the amounts of cash placed on the horses. This is a ***pari-mutuel*** betting process where the money is put into a pool. Odds fluctuate as additional bets are placed, right up to just before the race when they become fixed and odds and pay outs are calculated. Thus, Bill would not know his likely winnings until the start of the race. A portion of the bet total is taken by the bookmaker – this is the commission (approx. 15 % UK).

Take the 7 horses and assume that monies are placed on them for a win according to the table, after deduction of commission (at 15%). Assume actual total money placed was approx. £218 and after the deduction, £190 is spread across the horses.

The horse with most favourable odds is the 'favourite' ('Crosshair'), but if more bets are placed on horses nos. 6 and 7 ('Acehigh' and 'Last Oasis') before the start, then one of them may replace horse 3. Bill's calculations put 'Crosshair' at a better chance than the tote odds. If he feels that his information and subjective judgement are valid then he will gladly place his money on this horse at 3 to 1.

With this form of gambling, the money is sourced from the other gamblers, viz. if horse no. 2 wins and if it were a single individual who placed the £10 bet, they would receive the £10 plus £180, which originally belonged to the others in the pool. The form of the horses and other conditions are also reflected by this system – in reality, the majority of people who place the bets are knowledgeable – those 'in the know' put their money where their knowledge and study prompts them. They make take the advice of others more knowledgeable hence the 'tipsters' pages in racing newspapers and in online sites. If a particular horse has abilities well above all the other runners and that are secret then the whole betting structure falls apart and punters who have this knowledge can outsmart the bookies (Townsend).

Bill may not get 3 to 1 if more people place higher amounts on Crosshair.

Let's say a piece of information on a horse becomes available - suddenly more money goes on that horse - with fixed odds the bookie stands to lose a lot - but the pari-mutuel methods soaks up this – and the odds even out for that horse.

The 'win' bet is not the only wager that can be made and other possibilities include 'place' (1st or 2nd), 'show' (1, 2 or 3). Another type of bet is where the gambler places a wager on wins for more than one race (double, triple, etc.) enacted at the same horse track or at different ones. As would be envisaged, the probability for successive wins will fall and can be calculated by application of the multiplication and addition rules. In the UK, there is a bet called 'each-way' – a gambler chooses a horse to win and another to be first or second and even a wager that a horse will not win - a 'lay bet'. There are many variations of betting structure, according to those of different nations, countries and regions.

The analysis of the types of data in this chapter, as win:lose counts or more sophisticated measures, give probabilities that can be used to draw up betting odds. The quality of the results, i.e. their validity, depends in turn on the validity of the procedures used and the detail of the measures. Accuracy can be tested by the level of agreement between the predicted and actual in future play. This can be followed by revision and refinement of the procedures.

Chapter 16
Coins

16.1 Introduction

The previous chapter looked at some game probability aspects in a more general way. Following chapters will deal with specific randomising devices. As befits the first of these, we start with an action and an object, which in the latter case, although it has an obvious other function, is universally recognized as a randomising device: the coin and the coin flip.

16.2 Coin probability and coin events

Coin tossing, flipping or spinning as actions have played a role in simple games and probability problems since their beginning. Coins can be viewed as two-sided dice, although unlike dice, the two sides do not have the same general similarity in appearance. One side may often protrude more than the other may and weight distribution may not be exactly balanced between the two sides. Additionally, coins have a third 'side' made up by the rim. This is not of the same form and symmetry as that of the 'head' (H) and the 'tail' (T) sides. These characteristics support observations that the probability 'split' for a coin flip may not be '50:50'. Experiments have shown bias for one side over the other, especially when the coin is spun rather than tossed. Theoretically, there is also a chance that a coin flip onto a flat surface could result in an 'edge landing'. Some calculations have been done for this (Mosteller[a]). Chances would be higher for this event with 'thick' coins such as the current UK pound coin, reducing as the coin thins.

For probability trials with a coin, the randomisation is obtained via the toss or flip where the coin spins in the air. It can be allowed to fall onto any flat level surface or it is caught and slapped onto the surface or onto the back of the other hand of the flipper. This latter action would obviate an edge landing. It can also be spun directly on a flat surface and allowed to settle, which has been mooted as resulting in bias towards the head (Gelman, Haigh).

For the discourse herein, we will assume that the coin is thin, rendering edge landing negligible, and that H and T physical differences are insignificant, for example as with old worn coins; the coin is flipped and caught or it can be allowed to settle. Multiple coin probability experiments can involve the actual coins or one can be flipped repeatedly, thus an event with 'Z coins' can mean 'one coin with Z flips'. Assuming no sleight of hand and no tampering with the coin, then a random outcome will occur with probabilities of 50% for either outcome (a 'fair' coin).

Terminology for coin outcomes is relatively simple and was described in Book One, Chapter 1: we allocate H and T for the outcomes and indicate results for multiple coins as a sequence of these single coin outcome elements (e.g. TTHTHHTT) or as a count ('three Hs and five Ts').

16.2.1 Coin events

Probability problems with coins come under the headings of events with the formats described in Book One Section 5.3, in particular, those with specified *content* or in *sequence,* exactly so, or as a range or group of possible outcomes (Table 16.1). This does not exhaust the possibilities for coin outcomes and some of these events can be expressed in different ways, e.g. '*0 Hs'* with 6 coins, is also '*a run of 6 Ts'.*

16.3 Single coin probability

Each flip with a coin conforms to a ***Bernoulli trial*** (J. Bernoulli, 1654-1705) circumstance, where we have but two possible outcomes and any experiment of tossing a fair *single coin* has the 50:50 balance for the outcomes as seen in a simple diagram (Fig.16.1). The event outcome *'getting a head on a single toss of a coin'* shades half the area and the probability for a *'tail'* outcome is the same, with $P = {}^1/_2$ for each.

Table 16.1 Examples of coin probability events and related measures

Occurrence	Event ('probability of getting -')
Named outcome	A H on 1 coin
Named sequence	HTHT with 4 coins
	100 Hs (100 coins)
Named content	Two Ts and a H with 3 coins
	No (zero) Hs with 6 coins
	Any count of four Hs with 7 coins
Ranges & groups of outcomes	At least nine Ts on 15 tosses
	At most seven Hs on 21 tosses
	An even no. of Ts' with 10 coins
Named run	A run of four Hs with 10 coins
Higher probability (count or seq.)	Three Hs (5 coins) cf. three Hs (6 coins)
Expected value	No. of Ts on tossing 6 coins repeatedly
Expected no. of flips	To get a T on single coin
	To get a sequence of HHH with 5 coins
Expected no. of appearances	Of four Ts on flipping 9 coins 20 times
Named sequence in a game	HHH beating THH in Penney Ante

Single coin experiments with one trial are used as such to decide many issues, typically for 'who goes first' in sport games as covered in Chapter 15. Data from single coin experiments conform to a *uniform distribution* (Book One Section 8.2), i.e. there is equal probability for each outcome.

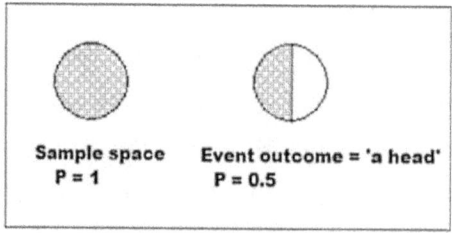

Figure 16.1 Diagram of a single coin flip

Using a single coin and tossing it once appears limited but repeating the trial can be viewed as a way of checking on bias (Book One Section 3.6) in a coin. Here we would expect on average a balance of Hs and Ts. This would be a long-term average and performing ten tosses would only be effective at detecting gross bias. A 100 toss experiment (Fig. 16.2) produced a plot of the results as a two-column chart, with a symmetrical form for the theoretically based counts, and a less so one for the empirical. There is a slight divergence from the unbiased state but this is still well within the variation one would see with an unbiased coin (see example in Section 16.5). Note that the manner of recording the results above does not take into account any sequence, and such a viewpoint would be taken as a multiple coin experiment.

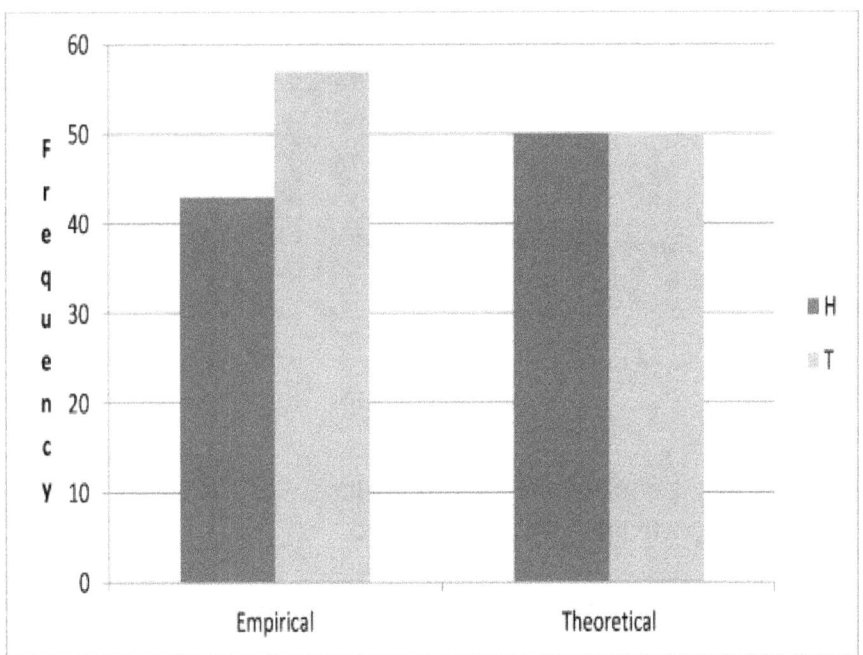

Figure 16.2 Empirical (100 Bernoulli trials) and theoretical distribution of single coin toss outcomes

16.4 Multiple coin probabilities

16.4.1 Multiple coin outcome distribution

With multiple coins (flips), the distribution of Hs and Ts conforms to that of the **binomial** or to the **binomial model** (Book One Section 9.2). Two or more coins are flipped and the frequency of H (or T) occurrences can be graphed to illustrate the distribution. Figure 16.3 shows the results of such an experiment where 10 coins were flipped 45 times and the occurrence of the number of Hs recorded.

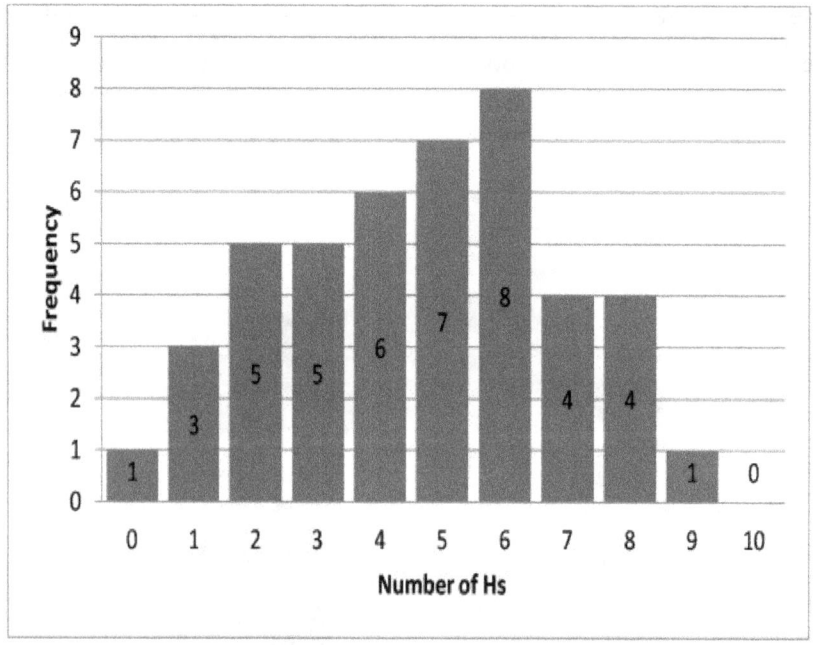

Figure 16.3 Distribution of outcomes(count of heads) for 10 coins flipped 45 times

The probability function for this distribution (the binomial) can be used to solve many coin problems where there is a fixed number of trials (see below).

16.4.2 Multiple coin events

Events for two or more coins have formats as described in previous sections, namely, sequences, contents, runs, etc. To enable probability calculations we need first to determine the *total number of outcomes* for any number of coins.

All possible outcomes

We can easily get this count for the outcomes via the basic counting principle (Book One Section 7.1.1) as,

No. of coin outcomes (sequences) = 2^n

Where, n = no. of coins or flips

In Excel: = 2^n $\qquad\qquad$ **(16.1)**

For example, 4 coins generate $2^4 = 2 * 2 * 2 * 2 = 16$ different sequences.

Tree diagrams (Book One Section 4.3, Fig. 4.8) are also invaluable in this respect, but they soon get large (Fig. 16.4). They have the advantage of showing the sequences and other aspects directly.

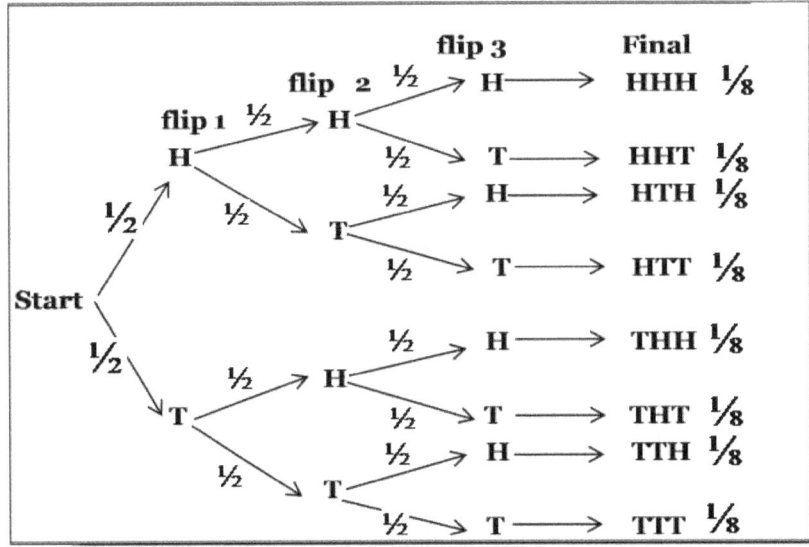

Figure 16.4. Tree diagram of outcomes in a 3 coin experiment

We can see that there are 8 outcomes in total, but only 1 with the sequence HHH, which is also a 'run of 3 Hs'.

Coin outcome generation

All possible outcome sequences for coins can be generated by looking at each one as a series of '0s' and '1s', e.g., HHTTHH is equivalent to '110011'. Some readers may recognize this as binary number (a number to the base 2 rather than the usual decimal base of 10). This need not concern us as we can generate all forms of a binary sequence for a fixed length in Excel using the DEC2BIN( ) function, for the number of sequences possible (Spreadsheet Table 16.1). The total number is calculated as above and a list prepared of this count, from 0 to the total minus 1, is placed on the leftmost column in consecutive order (col. A).

Enter 0 in cell A4, highlight the cell and place the cursor on the lower R-corner to turn on the Fill handle, then R-click this and select Fill Series and drag down to fill column A. The function statements follow along to the right, starting in cell B4. The example in the table can be extended and results tabulated into more than one column as required (maximum of 9 coins/flips in Excel® 2010 with these settings, but it is possible to extend the string length to cover 10 or more).

Using such tables for a manual search is limited as the number of entries increases, e.g. 10 flips has over a 1000 sequences (1024) – poring through such a list may be rewarding but can be tiring! However, the lists can searched with Find and Replace utilities and by use of the COUNTIF( ) function in Excel. The individual outcome on the right can be formed into a text string using CONCATENATE(E4,F4,G4,H4,I4) to aid this. Additionally, a tree diagram of even 5 coin flips is large and much more difficult to construct and we can use such tables for counting possible outcomes, with more confidence (see examples below).

Spreadsheet Table 16.1

COIN OUTCOME GENERATOR

	DATA: no. coins (n) =	5						
	no. sequences =	32 (= 2^n)						

A	Cell B4 = DEC2BIN(A4,n)	C4 = TEXT(B4,"0000000000")	D4 = RIGHT(C4,n)	E4 = IF(MID($D4,1,1) ="1","H","T")				
0 (A4)	(B4) 00000	0000000000	00000	T	T	T	T	T
1	00001	0000000001	00001	T	T	T	T	H
2	00010	0000000010	00010	T	T	T	H	T
3	00011	0000000011	00011	T	T	T	H	H *
4	00100	0000000100	00100	T	T	H	T	T
5	00101	0000000101	00101	T	T	H	T	H
6	00110	0000000110	00110	T	T	H	H	T *
7	00111	0000000111	00111	T	T	H	H	H 1
8	01000	0000001000	01000	T	H	T	T	T
9	01001	0000001001	01001	T	H	T	T	H
10	01010	0000001010	01010	T	H	T	H	T
11	01011	0000001011	01011	T	H	T	H	H *
12	01100	0000001100	01100	T	H	H	T	T *
13	01101	0000001101	01101	T	H	H	T	H *
14	01110	0000001110	01110	T	H	H	H	T 1
15	01111	0000001111	01111	T	H	H	H	H
16	10000	0000010000	10000	H	T	T	T	T
17	10001	0000010001	10001	H	T	T	T	H
18	10010	0000010010	10010	H	T	T	H	T
19	10011	0000010011	10011	H	T	T	H	H *
20	10100	0000010100	10100	H	T	H	T	T
21	10101	0000010101	10101	H	T	H	T	H
22	10110	0000010110	10110	H	T	H	H	T *
23	10111	0000010111	10111	H	T	H	H	H 1
24	11000	0000011000	11000	H	H	T	T	T *
25	11001	0000011001	11001	H	H	T	T	H *
26	11010	0000011010	11010	H	H	T	H	T
27	11011	0000011011	11011	H	H	T	H	H *
28	11100	0000011100	11100	H	H	H	T	T 1
29	11101	0000011101	11101	H	H	H	T	H 1
30	11110	0000011110	11110	H	H	H	H	T
31	11111	0000011111	11111	H	H	H	H	H

Coin outcome sequences

A *coin sequence* has the individual coin outcomes in order, e.g. 5 coins can have five single coin elements resulting in various 5-part outcomes such as HHTHH or TTTTH. Thus, a coin sequence is the ordered outcomes from tossing 2 or more coins. Each sequence is unique. For example, HHTHTT is a sequence, resulting from a 6 coin (flip) experiment, which differs from another possible sequence, TTTHHH, which has the same content. The probability for the sequence can be determined by use of the single coin(flip) probability, then stringing these individual probabilities together according to the intersection of events principle (Book One Section 5.2.1), i.e. the multiplication rule ('AND'). This is applicable because each flip is independent. For example, HTHT consists of four separate events, linked by AND:

> Getting a H on flip 1 (P = $^1/_2$)
>
> AND
>
> Getting a T on flip 2 (P = $^1/_2$)
>
> AND
>
> Getting a H on flip 3 (P = $^1/_2$)
>
> AND
>
> Getting a T on flip 4 (P = $^1/_2$)

Hence, P('*Getting sequence HTHT, 4 flips*') = $^1/_2 * {}^1/_2 * {}^1/_2 * {}^1/_2 = {}^1/_{16}$

Remember, this is an *ordered* outcome, as explained above, but it applies to any named sequence of fixed length and essentially the calculation reduces to the individual probability raised to the power n:

> **Probability of a coin sequence = $(^1/_2)^n$**
>
> **In Excel: = (1/2)^n**
>
> Where, n = no. of coins or flips   **(16.2)**

Each sequence in the sample space is of equal probability. Thus, getting HHHTT on five coin flips has a probability of $^1/_2{}^5$ and each of the other 31 sequences in such an experiment has the same probability. Perusal of tree diagrams and outcomes in Spreadsheet Table 16.1 will confirm this. Thus, for the general case, i.e. any number of coins/flips, each sequence is unique and occurs once in the distribution. For example, '*What is the probability of getting the sequence TTHHHTTHHT with 10 coins?*',

$$P = {}^1\!/_2{}^{10} = {}^1\!/_{1024} = 0.000976563 \approx 0.1\%$$

Each of the other 1023 outcome sequences has this probability. Coin games that use sequences include the *Penney Ante* game (below) and any coin game where players flip the coin in order, such as in a row or a circle.

Coin outcome content (counts of Hs and Ts)

A *coin content* event is stated with reference to what the outcome contains in terms of the number of heads or tails, e.g. '*four heads on tossing 6 coins*' – order is not considered and a result which has this content (any sequence) in terms of outcomes will qualify. This form of outcome is a combination in the sense described above. The event specification can include numbers of both Hs and Ts, or just one, with the other automatically implied. Games with coin content could include players tossing multiple coins and the winner(s) is the one with most Ts (or Hs).

Probability can be determined for any one of such sequences as above, then multiplied by the number of sequences possible (according to permutation formula (Book One Section 7.2.2),

Probability of a coin outcome content:

$$= (^1\!/_2)^{\,n} * n!/H!T!$$

In Excel: = (1/2)^n * FACT(n)/(FACT(H)*FACT(T))

or = (1/2)^n * MULTINOMIAL(H,T)

Where, n = no. of coins or flips

H, T = no. of heads and tails, respectively **(16.3a)**

(Readers may also notice that as H+T = n, n! / (H!T!) = ${}^n C_H$ or ${}^n C_T$ and the expression is equivalent to use of the binominal function (see below))

Thus, *'two Ts and a H'* has probability of $1/2^3 (= 1/8)$ as a single sequence, but it can occur in $3!/(2!*1!) = 3$, different forms: 'TTH, THT, HTT'. Thus,

$$P(\text{Content} = \text{'two Ts and a H, 3 flips'}) = 1/8 * 3 = 3/8 = 0.375 \ (38\%)$$

Similarly, P(Content = *'three Ts, four Hs, 7 flips'*)

$$= 1/2^7 * 7!/(4! * 3!) = {}^{35}/_{128} = 0.273438 \ (27\%) \ ({}^7C_4 = 35)$$

Another way to view this calculation is found when using combinatorics directly. This can be seen if we recall one way in which formula (7.5a) (Book One) can be used, i.e. as counting the number of ways for items to be fitted into slots. We can view the coin sequence length as the number of slots and the number of Ts or Hs as the items to be fitted in. Hence, 'two Ts on 5 coins' becomes 'slotting two Ts into two of five available locations': no. of ways = ${}^5C_2 = 10$ and the probability is ${}^5C_2 / 2^5 = {}^{10}/_{32}$. This is equivalent to the method above: $5!/(3! * 2!) * 1/2^5$.

We seem to be breaking a rule here, in that we are dividing combinations by permutations, i.e. the denominator above gives all the sequences for 5 coins – the numerator is calculated with a combination formula. Although this is so, we are getting a count of sequences as explained in Book One (see Sections 13.2.2 & Section 7.3.1, respectively).

Also, using Excel on the example above:

P(*'three Ts on 7'*) = COMBIN(7,3) / 2^7 = ${}^{35}/_{128}$

$$= (1/2)^7 * FACT(7)/(FACT(4)*FACT(3))$$

$$= (1/2)^7 * MULTINOMIAL(4,3) = {}^{35}/_{128} = 0.273438 \ (27\%)$$

(There is a multinomial calculator in Section 18.4 that can be used for coin events).

Binomial function and coin probability

The numerator in the latter calculation (COMBIN(7,3)) is the *binomial coefficient* and a more rapid method for coin problems when using spreadsheet or calculator is to invoke the ***binomial probability function*** (BINOM.DIST( ) in Excel**).** The coefficient determines the number of combinations of Hs (Ts) as content.

With coin probability, the function views the Hs and Ts as the user wishes, i.e. as 'successes' and 'fails' or vice versa. The details of the binomial calculation are given in chapter 9 (Book One Section 9.2). With coins, the necessary data in the arguments are taken as: the 'favourable' outcome of interest is identified as the 'success', and we rephrase the problem as 'the number of successes within the number of trials (flips)':

$$P_{coin\ content} = BINOM.DIST(k, n, 1/2, cumulative)$$

Where, k = no. of successes

n = no. of coins (flips)

cumulative = 0 or 1   ***(16.3b)***

The probability of each trial is the same (0.5) throughout the experiment, thus '*getting exactly two heads on three tosses*' becomes 'achieving 2 successes in 3 trials' and 'cumulative' is set = 0, then:

P = BINOM.DIST(2,3,0.5,0) = 0.375000 = $^3/_8$ (38%)

Provided we are not concerned with order (i.e. sequences are not considered), we can go on to calculate probabilities of outcome content for any number of coins and successes from zero or more, e.g. '*exactly zero Hs on 6 flips*',

P = BINOM.DIST(0,6,0.5,0) = 0.015625 (2%)

($\equiv$'*6 Ts on 6*' or '*a run of 6 Ts*')

'*Exactly one hundred 'Hs on 100 flips*',

P = BINOM.DIST (100,100,0.5,0)

0.0000000000000000000000000000001 ($\equiv$ '*zero Ts on 100 flips*')

A tail can also be viewed as the success: '*Exactly eight Ts on 9 flips*'

P = BINOM.DIST(8,9,0.5, 0) = 0.0175781 (2%)

This probability is the same as that for '*one H on 9 flips*' and the other examples can be converted to the 'tail equivalent'. These can all be confirmed using the alternative calculation with the combination 'slots' as:

P('*eight Ts on 9*') = COMBIN(9,8) / 2^9 = 0.0175781

(Also by multinomial calculator in Chapter 18 Section 18.4 - enter dice size as '2', no. dice as '9' and Hs and Ts as '1' and '2', respectively).

The probability of any sequence with a particular content can be calculated by dividing the binomial result by the number of possible sequences. Thus using the above,

P('*TTTTTTHTT*') = BINOM.DIST(8,9,0.5,0)/(n!/(1!*8!))

= BINOM.DIST(8,9,0.5,0) / MULTINOMIAL(1,8)

= (1/2)^9 = 0.00195313 (0.2%)

Ranges and inequalities in coin probability

For ranges and groups of outcomes we can simply sum the probabilities of the component parts viz. for '*3 or less Ts*' with 5 coins, we sum probabilities for 3, 2, 1 and 0 successes. However, it is easier to manipulate the binomial formula as explained in Book One Section 9.2. For coins, only the number of successes can be reset to cover a range, i.e. the probability of the event cannot be extended, as is the case with dice. Consequently, successes up to a specified maximum 'at most ' or at and beyond a minimum ('at least') can be included and the cumulative argument must be set to TRUE(1) to cause the probabilities to be summed, viz.:

'*At most seven Ts on 21 tosses*' by,

P = BINOM.DIST(7, 21,0.5,1) = 0.09462 (9%)

and '*At least nine Hs on 15 tosses*' by

$$P = 1\text{-BINOM.DIST}(9\text{-}1,15,0.5,1) = 0.3036 \ (30\%)$$

These settings are useful in some analyses, such as when testing claims regarding possible bias in a coin (see below under expected value). In other events, with a restricted group of outcomes, such as '*What is the probability of an even number of Ts with 10 coins*' the inequality formulae cannot be used. The content for 0,2,4,6, 8 and 10 Ts must be calculated separately and summed (most easily using Excel with the BINOM.DIST( ) and SUM( ) functions), although there is a small short-cut (Problems & Exercises).

The perception of H and T occurrences

The 0.5 probability of the coin toss action can be misinterpreted as a belief that on any number of tosses, a 50:50 balance of 'Hs' and 'Ts' has a high ('good') probability, or that such an event has a 50% chance. The binomial calculation can show clearly that, except for 2 coins, this is untrue - outcomes with equal numbers occur less frequently as the number of trials increase. A few examples, with even numbers of tosses, show this:

$$P('H{=}T, \ n{=}4') = \text{BINOM.DIST}(2,4,0.5,0) = 0.375000 \ (38\%),$$

8 coins (27%), 16 coins (20%) and 10% for 64 coins.

The probability of getting an equal number of heads and tails falls as n increases and it soon becomes 'poor', eventually becoming a rare event (< 1%). However, it will still be the event with the *highest probability*, it's just that there are many other events possible and these increase rapidly with the number of coins. The above result contrasts with the overall *proportion* of Hs, which is calculated by summing the accumulated number of Hs after each toss and dividing by the number of tosses. If, in an experiment with 64 tosses, 36 Hs have accumulated, then the proportion is $^{36}/_{64} = 0.5625$ (56%). This contrasts with the probability of getting H = T (P = 10%). Doing this calculation for many tosses shows that the relative frequency of the occurrence of 'Hs' gradually approaches the population proportion (0.5). By

100 flips and more, the proportion is very close to, or at, 50%.

This may seem contradictory and counter-intuitive. It comes down to the distinction between these terms. The proportion in this context does not refer to the number of outcomes that can result in 'equal Hs and Ts'. Take the case for 16 coins - there are 12870 of these, out of a total of 65536 where H = T:

$$\text{Proportion} = {}^{64}/_{128} = 0.5, \text{ but } P('H{=}T') = {}^{12870}/_{65536} = 0.196381 \ (20\%)$$

Therefore, in about 1 in 5 tosses of 16 coins you would get exactly eight 'Hs'. A simulation of 100 repeats of 16 coin tosses might get about 20-25 with H=T, but the overall proportion of Hs is 0.49 to 0.51.

Thus, games based on coin tossing can exploit the fact that a 50% probability for H=T applies in but one case: 2 coins. Also useful as previously stated, is that it is the most probable content outcome in the distribution for any even number of coins. This leads to another point of interest: *'What is the most probable number of heads on tossing Z coins?'*

Most probable number of Hs (or Ts) = Z/2
where, Z = any even no. of coins

Thus, with 18 tosses P(*'nine Hs'*) = 0.185471 (19%), which has higher probability than any other exact content event. This is the *mode of the binomial distribution* (formula (9.1b) Book One). For uneven numbers of coins the distribution is bimodal. Consequently, for 19 flips the counts either side of 9.5, namely, 'nine and ten Hs' are the most probable counts. This effect results in some coin distributions having the same mode, as with 'three Hs on 5' (modes 2 and 3) and 'three Hs on 6' (mode 3). The uninitiated would assume that that 6 coins would have the higher probability but the actual results are again counter-intuitive, as the events have the same highest probability (0.3125) due to the bimodal state for one. Sequences with three Hs occur in numbers of 5!/(3!*2!) and 6!/(3!*3!) and also have same the probabilities (${}^{10}/_{32}$ and ${}^{20}/_{64}$, respectively). However, we cannot compare specific sequences, such as in the case of HHHTT cf. HHHTTT, where P = $^1/_{32}$ and $^1/_{64}$ respectively.

Coin sequence runs

Many other formats of coin results are possible leading into some very 'deep' probability problems. One such is the **run**, which appears in various forms (Section 15.2.3 & Book One Section 5.3.3). For coins, we have the 'isolated' run sequence, where the probability is simply the product of the individual coin ones and is calculated using formula (16.2) for a coin sequence (P = $^1/_2$ "), where n is the number of coins and also the length of the run. Hence:

P('*four Ts on 4 tosses*') = $^1/_2$ ^{4} = $^1/_{16}$

P('*getting 20 Hs in a row*') = $^1/_2$ ^{20} ≡ HHHHHHHHHHHHHHHHHHHH

Another form of this event, the 'simple mixed run sequence', has successes and fails, but all the successes (fails) are in the run, e.g. – a run of '6 Hs in 11' and no other Hs occur. Probability is that for the event as one sequence, times the number of ways it can occur and is given by formula (5.3) (Book One Section 5.3.3), modified for coins as:

P(simple run with coins) = (f + 1) * 0.5^n

where, n = no. coins or flips   f = no. fails (length of run r = n - f)

Thus, for P('*(HHH) on 5*'), n = 5, f = 5-3 = 2 and,

P = (2+1) * 0.5^5 = $^3/_{32}$

For a run of '6 in 11',

P('*TT(HHHHHH)TTT*') = (5 +1) * 0.5^11

The most complex type of coin run sequence structure is where the run is accompanied by mixtures of Hs and Ts and sometimes of other runs, e.g. T(HHHHH)THT(HH), which has a '5 H' run and a '2 H' run. As explained in previous chapters, calculation is also complex, but the probability for these and more complicated run structures can also be determined by other methods.

Thus, tables produced by the generator above can be used. Each occurrence in the list is marked with a '1' opposite, then these can be added up using the SUM( ) function. We can see in Spreadsheet Table 16.1 that simple and more complex runs of three 'Hs' have been marked in this way: it occurs 5 times for a sequence with a run plus mixtures of Hs and Ts, such as '(HHH)TH':

$$P(\textit{'complex run of 3 Hs, 5 flips'}) = {}^5/_{32}$$

We can see that this probability is higher than that for the simple run above, under the same settings. We could also extend this event to *'a run of at least 3 Hs'* and $P = {}^8/_{32}$ (with two runs of four and one of five added in). Simulation of such event outcomes is possible as well, followed by counting, which can be performed using the methods in Book One Section 14.3.1 such as that applied for the *Penney Ante* example (Simulation 3). Some examples of complex run probability estimates with simulation (1,000,000 trials) are:

$$P(\textit{'at least length 4' 10 flips}) = 0.24531 \, (25\%) \, (\text{correct } P = 0.247070)$$
$$P(\textit{'1 or 2 of length 4' 10 flips}) = 0. \, 0.137557 \, (14\%)$$
$$(\text{correct } P = 0.137695)$$

With more coins, runs can get longer if there is more room in the sequence,

$$P(\textit{'20 flips, run of 5 or more'}) = 0.249 \, (25\%)$$

With very large numbers of coin flips it is possible to approximate the run probability by using the Poisson function (Section 15.2.3), so, $r = 15$ and $n = 100$ gives lambda $= 0.5^{15} * (1 * (100 - 15) * 0.5 = 0.001327515$, then,

$$P(\textit{'100 COINS, run of 15'}) = \text{POISSON.DIST}(1,0.001327515,0)$$
$$= 0.00132575 \, (0.13\%)$$

For the run length that is most probable, we can look at diagrams and tables and count occurrences, but logic suggests that the occurrence of longer runs is less frequent in the sample space. Hence, for 2 or more flips, a single run of length 2 (HH) is the most probable. As run length decreases run probability for a single instance falls but '1 or more' probability rises; as run length increases probabilities converge as there is less room for more than 1 of the larger runs. In 100 coins '1' and '1 or more' at 20 length are very close and thereafter virtually identical.

These probability values could be used for games, based on the nature of run obtained. We can also look at runs in the win:lose balance for coin games, as in 'how many times have you lost in a row at that Penney Ante game?' If the player is not aware of the strategy for this game (see below), their losing streak is likely to be a long one.

16.5 Expected value and variability for number of Hs (Ts)

The random variable under consideration is designated as the number of Hs or Ts, which takes values from zero up to 'n', the number of coins[2]. To get the average number and its degree of variation, we use the mean and variance of the **binomial distribution** (Book One Section 9.2), namely n*p and n*p*(1-p), respectively, where n is the number of trials and p = 0.5, for coins.

For example, '*What is the expected no. of Ts on 6 coin flips?*' The random variable (X) is the number of tails and n = 6,

$$E(X) \text{ 'no. Ts, 6 flips'} = n*p = 6 * 0.5 = 3$$

We would expect to get, on average, three 'Ts' with 6 coin flips. The mode is also 3 at 31% probability. The variability of such an experiment would be,

$$V(X) = n*p*(1-p) = 3 * 0.5 = 1.5 \qquad SD = \sqrt{1.5} = 1.224745 \approx 1$$

---

[2] For a single flip, the random variable can take but two values, 0 or 1, as is the case with all 'single selection' type experiments

Thus, on average three Ts, ranging from 2 to 4.

The use of E(X) as an indicator of bias was described in Book One Section 3.6 with a coin example of 100 tosses with a result of 65 Hs. The E(X) for such an experiment is,

E(X) 'no. Hs, 100 flips' = 100 * 0.5 = 50

V(X) = 100 * 0.5 * 0.5 = 25   SD = 5

Therefore, there is a divergence, as we would expect 45 to 55 'Hs'. The probability for *'65 or more 'Hs'*,

P = 1 - BINOM.DIST(65-1,100,0.5,1) = 0.00175882 (0.18%)

This does appear to be a rare event and a statistical test shows that the divergence from the expected value is significant – the coin appears to be biased, favouring 'Hs' over 'Ts' and any count of 'H' over 59 become significant for the same experiment.

The methods can be extended with the binomial parameters to specific numbers of repeats (Book One Section 9.2), e.g. to get the average no. of appearances of 'four Ts' on tossing 9 coins, 8 times:

P = BINOM.DIST(4,9,0.5,0) = 0.25

n = 8 and E(X) = n*p = 8 * 0.25 = 2

Outcomes containing 'four Ts' would appear 2 times out of 8, on average.

Average number of flips and success by the nth trial

Other distributional functions and parameters can assist with other coin problems, in particular 'how long before' problems and others. If it is stated that the experiment involves waiting for an occurrence then we have a possible infinite number of trials. For example, when tossing a coin, *'how long* (i.e. how many tosses) *until I get a head?'* As explained in chapter 10, this could involve a wait to infinity, but the probability gets lower and lower as the trials progress, although some believe the possibility should increase (counter-intuitive). Instead of the number of Hs or Ts the random variable is 'the number of flips' to achieve the particular event. This can take values of 1 flip, 2 flips, 3 flips ... ∞, but we require the average number of flips. We can deal with some forms of this probability problem by use of the **geometric** and **negative binomial distribution** models (Book One Chapter 10). The parameters of these distributions can tell us how long on average we need to wait for success. So, we can calculate the average 'wait' and also the probability of one or more success on any number (no limit except infinity) of trials.

A fair coin is flipped until a 'T' appears – *'how long will it take?'* To answer this we must use the parameters of the geometric distribution, which applies for a single success. For the geometric:

$$E(X)_{geometric} = 1/p \qquad V(X)_{geometric} = (1-p)/p^2$$

'X' is the random variable = no. of trials until success, and p is the probability of success

When flipping a coin, X can take a value of one of 1, 2, 3, 4, etc., but on average it will be the expected value of the distribution. So for $p = {}^1\!/_2$ as above,

$$E(X) \text{ 'no. trials, flip 1 coin until one T'} = 1/{}^1\!/_2 = 2$$

$$V(X) = (1-{}^1\!/_2)/({}^1\!/_2)^2 = 2 \text{ and } SD = \sqrt{2} = 1.414214 \approx 1$$

On average, success (a tail) will occur on the second trial, ± 1.

Two flips may achieve the event but what is the probability of it being the 1st, 2nd or 3rd ...? *'What are the chances of having to wait until my 2nd flip of a coin to get a H?'* The most convenient method of calculation for this is by use of Excel's **NEGBINOM.DIST( )** function. We are looking for a success on the 2nd trial, there is 1 fail, the probability for a success is ${}^1\!/_2$ and we want the exact probability:

P(*a H on 2nd trial*) = **NEGBINOM.DIST(1,1,1/2,0)** = 0.250000 (25%)

Similarly, on 1st = 50%, on 3rd = 12.5%, .. on 10th = 0.1% ...

This illustrates the 'decreasing probability' nature of the geometric distribution (with these circumstances).

Repeats of other experiments (simple or compound) can be accommodated; all that is needed is the probability of the event. Take the experiment of tossing 5 coins - *'what is the average no. tosses of 5 coins to get a content of HHH?'* Using formula (16.3a),

$P = {}^{10}/_{32}$ and

$E(X) = 1/ {}^{10}/_{32} = 1 * {}^{32}/_{10} = 3.2 ≈ 3$

Likewise, the average no. throws of 5 coins to get *'a run of at least 3 Ts'*,

$P = {}^{8}/_{32}$; $E(X) = 1/ {}^{8}/_{32} = 1 * {}^{32}/_{8} = 4$ (coins tossed in order each time)

So, when tossing 5 coins one after the other in a line, on average, 4 goes would be required before you have a row of 3 or more Ts. (Note that these results differ from a ' single coin tossed until you get 3 Hs', or 'a run of 3 Ts'), for which see below).

For multiple successes, as when the same coin is flipped until three 'Hs' have appeared, we can apply the negative binomial parameters (6.7a) in a similar manner:

$E(X)_{neg.binomial} = k/p$ $\qquad$ $V(X)_{neg.binomial} = k * (1 - p)/ p^2$

where, k = no. of successes $\quad$ p = probability of success in individual trial

Consequently for the example, there are 3 successes and k = 3 and on average,

$E(X)$ *'1 coin flipped until three Hs'* $= k/p = 3/0.5 = 6$

$V(X) = 3 * (0.5)/ {}^{1}/_{2}{}^{2} = 6$ $\quad$ SD ≈ 2

In the long term, it should take 6 flips, but the deviation is ± 2.

For the probabilities at various points, the **NEGBINOM.DIST( )** function is used with the data for 0 or more fails and 2 or more successes. So, for the above, the probability of 'not getting' the event until at the average (6 tosses), is given by the function with 3 failures and 3 successes:

P('*3 Hs by 6th trial*') = NEGBINOM.DIST(3,3,1/2,0) = 0.156250(16%)

By 3rd = 0.125000 7th = $^{15}/_{128}$ = 0.117188

Other problems cannot be solved by the above approach, e.g. in some of the above examples, the multiple successes (the 'Hs') are not necessarily consecutive. If we wish to know the average 'wait' for successive occurrences of Hs (or Ts) the above formulae would be invalid. For example, 'How many times on average do I need to toss a coin before I get 3 CONSECUTIVE Hs'? This is, of course a run of 'Hs' but we cannot use the run formulae described above for an expectation. It has been shown (Epstein) that the expectation of a run with coins is,

E(X) 'no. of flips, run length n, n coins' = $2^{n+1} - 2$

So, to get a run of '2 Hs' = 8-2 = 6 tosses, '3 Hs' = 16 -2 = 14, '4 Hs' = 32 -2 = 30.

The count for this type of event rises sharply and for '10 Hs' = 2^{11} -2 = 2046.

Tree diagrams can help elucidate, but such puzzles involving a sequence development over repeated flips can be complex and have counter-intuitive properties. One case was examined by simulation of the *Penney Ante game* (Book One Section 14.3.1) and below, with reference to Figure 16.4.

16.6 Coin games

Perhaps fewer games rely on coin tossing as the focus compared with the large variety for dice and cards. The simplest game is 'heads and tails'. Each player bets and a coin is tossed with 'a H' nominated as the winning outcome or they can call – 'Hs I win, Ts you win' etc. Odds are 1:1 so in a fair game winning should give the bet plus the stake. If the players take a preliminary toss to see who starts, then whoever goes first in this has a big advantage ($^2/_3$ vs. $^1/_3$; Section 15.2.3), so they should take it simultaneously. If the game itself uses one coin then order does not apply.

This type of game can be extended to more coins with consequent changes in probabilities for various outcomes. Two coins are flipped on a special paddle in the Australian game 'two up' (en.wikipedia.org) with odds based on the four outcomes.

Another game is where three players toss a coin simultaneously – 'odd one out' wins and takes all the coins. The probability for any one player being 'odd person out' can be seen by looking at a list of the outcomes by coin:

Coin (player1-p1)	p2	p3	Winner
H	H	H	none
T	H	H	p1
T	T	H	p3
H	T	H	p2
H	H	T	p3
T	H	T	p2
H	T	T	p1
T	T	T	none

Therefore, each player has a 2/8 (25%) chance of a win. When no one wins, the coins could rollover into the pot for the next winner. This game has been mentioned in some sources as a confidence trick scenario, with two of the players being in collusion (http://scams.wikispaces.com). As with other games, if the order of tossing is included, the odds change, as explained above.

Sequences come into a game where a coin is flipped in order introducing complexity in probability calculations. Four players (ABCD) flip a coin, one after the other. The first to get a H wins. On a <u>single round of one flip each</u> we have: outcome H (A wins $p = ^1/_2$), TH (B wins, $p = ^1/_2{}^2 = 0.25$) TTH (P3 wins $\frac{1}{2}^3$) and TTTH (P4 wins $\frac{1}{2}^4$). However, a tie could occur with TTTT and a second turn would commence and so on. Examining the outcomes, we can see that we have a geometric series. Using the procedure enacted in Section 15.2.3, we can terminate this to get the exact probability, using (formula (10.1c) (Book One)):

<u>A to win</u>

$$H = \frac{1}{2}$$
$$TTTTH = \frac{1}{16} * \frac{1}{2}$$
$$TTTTTTTTH = \frac{1}{16} * \frac{1}{16} * \frac{1}{2}$$
$$\text{common ratio} = \frac{1}{16} \quad \text{1st term} = \frac{1}{2}, \qquad P = \frac{1}{2}/(1 - \frac{1}{16}) = \frac{1}{2} * \frac{16}{15} = \frac{8}{15}$$

Probabilities for the other players can be calculated in a similar way and reveal the advantage of order (see Problems & Exercises).

Games can also be based on runs of different length and type using the probabilities explained above. With 5 coins we almost have 'coin roulette' as there are 32 sequences. Betting can be done on any set-up with coins for various events to get a range of odds, such as:

Events with range of odds on 5 coin flips	Odds
A T first	1 to 1
2H (3Ts) any order	2 to 1
At least 3 Hs in a row	3 to 1
4H	5 to 1
Run of 3 Hs by 2 Ts	9 to 1
5H	30 to 1

The *Penney Ante* coin game is singular in that it causes surprize in respect of the counter-intuitive probabilities, which are complex to elucidate by direct calculation. The game starts by each player choosing a sequence of length two (or more) of Hs, Ts of any mix. Play progresses by tossing of a single coin and outcomes are noted. Whichever sequence appears first, on the outcomes at any point, wins. The game is described in Book One Section

14.3.1 (Simulation 3) and probabilities were estimated using computer simulation. With a sequence of length two, the game is fair. On length three, the second player can always chose a sequence that has a higher probability of occurring than that of the first. It pays to be magnanimous and let your fellow-player go first in this game! Looking at Fig. 16.4 above, we can see that two players with sequences HHH and THH would have an equal chance on the 3rd flip. By the next flip, THH has two chances but HHH has none and this lead increases. Full lists of the odds are available in several texts (Nishiyama and Humble).

Many other coin games and puzzles exist where the coin is used more as a counter, as in positioning in layouts, e.g. 'shove half penny' and in the penny rolling game seen in fetes. Here a coin is tossed or rolled down a chute onto a grid marked out in squares (Mosteller[a]). The coin must land within the edges of any square to win a prize. Each square is marked with a prize name or value. With fun games at fund raiser events these will range from the booby' prize (0) to a jackpot (£1), with more numerous prizes of 2p, 4p, 10p and 50p scattered around randomly. Probability wise, although there are many squares we need only consider one. Within the square, the target area is limited to where the centre of the coin can land and still not touch the lines. This action can be likened to a random dart throw at the bull with the probability based on the area of the bull cf. area of the board. In this case, a hit is when the centre of the coin lands on this small interior area. To get the numbers, establish the dimensions of the coin and the square and its area then work out the area of the target:

Main square = 2 units x 2 units, area= 4 sq. unit

Coin = 1 unit diameter (e.g. the UK 2p coin has a diameter of approx. 1 in.).

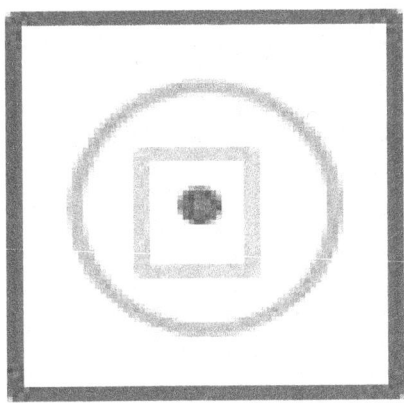

When the coin is exactly on the target there is a gap where it can move towards the lines of $^1/_2$ unit all the way round, so the interior target can be taken as a 1x 1square = 1 sq. unit area, hence,

$P = {}^1/_4 (25\%)$

If the square area is different in relation to 1 unit diameter coin:

Larger square of 3 x 3 = 9 sq. units, target sq. = 4 sq., $P = {}^4/_9$.

Smaller square of 1.5, area = 1.5^2 sq. units, target sq. = .5^2 sq., $P = {}^1/_9$.

Chapter 17
More round objects

17.1 Introduction

This application chapter looks at probability with a variety of objects. Most share a rough similarity with the coin in that they are all 'round' but in different ways: beans, balls, beads, marbles, icons and wheels, cylinders and circles. Some of the latter types are randomising devices themselves, others are manipulated by random selection. To begin, we look at a model of selection that appears in many if not all reference texts for probability.

17.2 'Beans in a jar' probability problems

Many probability problems have the *'beans in a jar'* format, where one or more objects are randomly selected from a source or pool. Puzzles with 'balls in an urn', 'jelly beans in a jar', 'beads in a bag', and 'marbles in a box', etc. abound in probability texts. These objects are used as explanatory tools because of their ease of access and their round shape, which ensures easy randomisation of a collection. True identity and location is concealed by an opaque cover in the form of the bag, jar, box etc. They provide simple examples of the application of combination and permutation formulae, but are applied in real games, some with big prizes. The basic scenario applies to selection not only of our round objects but to other game objects, cards in particular, and to any conceivable group of objects, such as icons, symbols, people, cars, defective items, etc. The items, categorised according to some feature that allows differentiation once revealed, are subjected to a random selection(s), for which probabilities can be calculated. Table 17.1 lists some examples.

Objects are physically removed in some of these examples but they can be replaced after each selection giving a different slant to the problem. The classical method can be used or calculation can be based on a probability distribution. When a selection is 'without replacement', the *hypergeometric distribution* model is applicable; for the *'with replacement'* case, the *multinomial distribution* is appropriate. In certain circumstances, it is also possible to use the *binomial distribution* function (Book One Sections 9.4, 11.2 & 9.2, respectively).

Table 17.1 Examples of 'beans in a jar' events

Occurrence	Event ('probability of getting -')
Choose 1 from several distinct	Get the 'Bank error in your favour' card® in Monopoly®
Choose 1 from several mixed	1 violet bean from a mix of 2 violet & 3 orange, in any order
	Draw a letter 'Q' from the letter bag in Scrabble®
Choose >1 from several mixed	Get 3 jacks on a deal of 3 cards
In order without replacement	Choose a yellow (5), a purple (13) and an orange (7) in order from a bag of coloured badges
Any order without replacement	Get 6 numbers in the lottery
	Get a 'house' in bingo
	Get a winning set (3 out of 8) on a scratch card
Get a higher probability	Two pair cf. three of a kind in poker
Expected no. of	Cuts, to get a 'K' on cutting single cards
	Draws, to get a sequence of double 1, double 2 and double 3 when drawing dominos

Monopoly® and Scrabble® ©Hasbro Inc.

Many probability problems can be made a lot easier by reducing the details to one of the above forms, such as choice of people from a pool, selection of lottery and raffle tickets, dealing cards, etc.

Calculation of 'beans in a jar' probability

Provided that the objects in this type of problem are clearly defined, the calculation can be relatively simple. Total and selected numbers of objects are specified along with the number of each type. All the objects together form a mixture of two or more distinct forms, those of the same type being indistinguishable, e.g. 2 red and 4 black, or all can be distinct, e.g. 99 tickets each with a different number. Events for selections of single and multiple of one type and multiples of different types can be formulated. The simplest case is where one object is drawn from the mixture. Selections with the same configuration nature and event structure have the same probability, i.e. P(*'2 from 2 red and 4 black, producing 1 of each'*) is the same as 2 selections from 2 purple and 4 orange, etc.

17.3 Single object selection

The simplest case of all is where each object is distinct as in a deck of 52 cards or a collection of 8 *snooker* balls, one of each colour. The probability for a random selection of any named item is,

$$P_{one\ selection,\ all\ objects\ distinct} = 1/N$$

Where, N = total no. of objects   *(17.1a)*

('N' is used rather than 'n' to conform with later formulae)

Thus, using (17.1a) for the examples above:

P(*'draw 1 card = 5♥'*) = $^1/_{52}$ P(*'pick the yellow ball'*) = $^1/_8$

A board game example: in *Monopoly®* © Hasbro Inc., there are 16 COMMUNITY CHEST ® cards, one is 'Bank error in your favour Collect £200'®. Thus, the probability of drawing this one at first choice from the pack is P = $^1/_{16}$.

With mixtures of different forms each with the same or different counts, the problem gets more complicated, even with still selecting 1 object. For example, '*I have a small jar containing 3 orange and 2 violet coloured beans*

– I select 1 at random – what is the probability of getting an orange bean? Getting a violet bean?' Classical probability provides ready answers to this as we can easily count the event subset and the sample space. The latter is the denominator in the classical formula (Book One (3.1)). It is made up by the number of ways of selecting 1 from N (the total number of objects) – with the coloured beans, we can see that this is equal to 5. The numerator is made up by the event subset provided by the number of ways of selecting 1 of the particular subgroup of interest, thus for *'getting a orange bean'* by a single selection can be achieved in 3 ways. To summarise,

P**_one selection, mixed objects_**

= (ways to select 1 from subgroup) /(ways to select 1 from N)

Where N = total no. of objects   **(17.1b)**

Hence, P(*'getting an orange bean'*) = $^3/_5$

Similar reasoning gives $^2/_5$ for violet.

Combinatorial formula can also be used, as:

Ways of selecting 1 from 5 = $^5C_1$ = 5 and,

Ways of selecting 1 of three = $^3C_1$

P(*'getting an orange'*) = $^3C_1 / ^5C_1 = ^3/_5$

Examination of the calculations above will show that we have in fact applied the **hypergeometric** distributional probability function (Book One Section 9.4). This applies as there is one overall main group of objects with a single separate subgroup of interest. The main group is identified as N (5 beans) and the subgroup of interest M (the 3 *orange* beans) and there is a selection (n) of one bean. The probability concerns the number of successes (m; or x in some texts) in this case 1, on a number of trials (n). The event is *'getting 1 orange on 1 selection'* in the action. Applying the function for this 'two group' case, N=5, M =3, m=1, n =1, thus,

P('*1 orange from 5 of 3 orange +2 violet*') =

$$= \frac{\binom{3}{1} * \binom{5-3}{1-1}}{\binom{5}{1}} = \frac{\binom{3}{1} * \binom{2}{0}}{\binom{5}{1}} = {}^{(3 * 1)}/_5 = 0.6$$

We also have the Excel spreadsheet function for this distribution and for the single object selection,

P('*1 orange from 3 orange, 2 violet on select 1*')

$= \text{HYPGEOM.DIST}(1,1,3,5,0) = 0.6 \; (^3/_5)$

This procedure applies to a single selection from any number of mixed objects. Provided the count of the one of interest and the overall total is known, we can get the probability of selection of one of that type. As explained in Chapter 9 (Book One), the identity of the 'remaining' objects are not necessary to the calculation and a problem of this nature was stated but not solved: a bag contains 99 mixed snooker balls with several different colours, where it is known that there are 8 green balls. The question '*What's the probability of making 1 selection and getting a green?*' was posed. Subtracting 8 from 99 leaves 91 balls in the 'rest' but their identity can be ignored, and:

$P = {}^8C_1 / {}^{99}C_1 = {}^8/_{99} = 0.08081$ (8%), confirmed by

$\quad \text{HYPGEOM.DIST}(1,1,8,99,0) = 0.08081 \; (^8/_{99})$

Single card selections of suit and rank are accommodated by this calculation. For example, *'drawing 1 card and getting a diamond'*. We view the pack as a collection of objects – 13 are diamonds and there are 52 in total, so getting a diamond on 1 draw:

$$P = {}^{13}C_1 / {}^{52}C_1 = 13/52 = {}^1/_4 \text{ and}$$

$$\text{HYPGEOM.DIST}(1,1,13,52,0) = 0.25 \, ({}^1/_4)$$

In addition, as there is only 1 selection the binomial can be applied to all the above examples,

$$P('1 \text{ orange on } 1') = \text{BINOM.DIST}(1,1,3/5,0) = 0.6 \, ({}^3/_5)$$

and other probabilities as P(*'1 green on 1'*) = 0.08081 ($^8/_{99}$) and P(*'1 diamond on 1'*) = 0.25 ($^1/_4$) (if objects are replaced, the function can be applied for more than 1 selection).

Expected value of single selections

The long-term average of a single selection is given by the expected value for the hypergeometric distribution: $E(X) = n * M / N$. For the beans, the random variable can take the values 0 (e.g. no orange coloured beans appear on the draw) or 1, but on average,

$$E(X) \text{ 'no. of orange'} = 1 * {}^3/_5 = 0.6 \, (\approx 1)$$

That is, in the long term, an orange bean would appear once or 3 times in every 5 single selections. For a single selection, the binomial expected value (n*p) also applies and confirms the above result.

Average number of draws and success by the nth trial

We can calculate the average count of selections and the probability for success at any stage during selections with other distributional functions as described above for coins. As there is a single selection, we can only do this for *with replacement*, i.e. *without replacement* becomes a multiple selection. The geometric function is appropriate for one success, as illustrated above for coins (Section 16.3). For the orange beans,

$$E(X) = n/p = 1/\,^3/_5 = 1.7$$

So on average, it should take between 1 and 2 draws to get an orange bean, returning selections after each draw. For a violet bean we would expect more and this is so,

$$E(X) = 1/\,^2/_5 = 2.5, \text{ hence 2 to 3 draws.}$$

In a game of Scrabble®, *'How many draws of letters will I need until I get a letter 'Q'?'* Assume 100 tiles and one 'Q' so on the first selection,

$$P('1st\ pick = Q') = {}^1/_{100}$$

and average no. picks $= 1/p = 100$

(selected letters must be replaced between draws)

In contrast, the letter 'E' (12 occurrences), on average requires only 8 selections,

$$E(X) = 1/{}^{12}/_{100} \approx 8$$

If tiles are not replaced the calculation is more complex (Problems & Exercises). The procedure is explained more in Sections 17.4 & 22.3.

17.4 Multiple object selection

Selection of more than one object raises new issues, in respect of *replacement* or not and *order* of choice. Some of the above procedures are applicable if the objects are retained as they are drawn and order is not considered, i.e. the hypergeometric model applies to these circumstances. Using the function gives values for 'any order', and specific order is illustrated later. Again, if all objects are distinct then it is only possible to get one of each type in any multiple selection, e.g. with lottery and bingo balls.

Without replacement and any order

A problem of this nature was described briefly in Book One Section 9.4. As a fuller account, take the example of the mix of 5 coloured beans above, i.e. orange (3) and violet (2) but this time a selection of multiple objects takes place.

If we constrain the multiple selection to events within one subgroup, then we can use the hypergeometric function up to the limit of that subgroup - so 'pick 3 and get 3 orange' is possible but 'pick 3 and get 3 violet' is not but 'get 2' is:

P('*pick 3 and get 3 orange*') = HYPGEOM.DIST(3,3,3,5,0) = 0.1

P('*pick 3 and get 3 violet)*' = HYPGEOM.DIST(3,3,2,5,0) = 0

P('*pick 2 and get 2 violet*') = HYPGEOM.DIST(2,2,2,5,0) = 0.1 (10%)

Reasoning by the classical approach, we can also use combinatorial formulae for the '*unordered, without replacement*' counts in such problems. Thus for the first example above, the sample space is 'the number of ways to choose 3 from 5' = $^5C_3$ = COMBIN(5,3) = 10. The event can be viewed as 'selecting 3 from 3 of orange = $^3C_3$ = 1 and

$P = {}^1/_{10}$ (10% as above)

Thus, an unordered selection, *without replacement*, has a probability of 10% (the ordered state is illuminated below). This type of calculation is readily applied to simple card probabilities,

P('*three Js on a deal of 3*')

= HYPGEOM.DIST(3,3,4,52,0) = 0.000180995 (0.02%)

More than one subgroup in the mix

When there are two or more subgroups within the main mixture of objects, we cannot apply the basic hypergeometric formula and the general version, the ***multi-hypergeometric*** (Book One Section 11.3) is required. Consider a larger mix of coloured objects, 10 balls consisting of red (1), white (2) and blue (7) and selection of 3. Using the classical method, the sample space for this problem is larger: Selection (unordered) of 3 from 10 = $^{10}C_3$ = COMBIN(10,3) = 120. Take as the question '*How many ways can I select 3 balls from 10 such that I get 1 of each colour?*' This would involve a selection of 1 red from a pool of 1, 1 white from 2 and 1 blue from 7:

$$= {}^1C_1 * {}^2C_1 * {}^7C_1 = 14 \text{ and } P = {}^{14}/_{120} = {}^7/_{60}$$

This is again the series of choices from three different groups as shown above and in Book One Section 7.4, but applying the general function (formula 11.2b) shows the same calculation:

$$P = \frac{\binom{1}{1} * \binom{2}{1} * \binom{7}{1}}{\binom{10}{3}} = 0.116667 \; ({}^7/_{60}, \; 12\%)$$

As explained in Book One, we do not have an Excel function for this but it is easy enacted using the Excel version of formula (11.2b):

P = (COMBIN(1,1) * COMBIN(2,1) * COMBIN(7,1)) / COMBIN(10,3)

= 0.116667

Additional problems follow in Chapters 21 and 22 (cards), illustrating more features and scope with the hypergeometric formula.

In order, without replacement specification

The above multi-object selections were for the 'any order' state. Probability for a specified order can be obtained by first calculating the number of possible arrangements of the resulting selection. This is done using permutation formulae (Book One Sections 7.2.1 & 7.2.2).

In a fantasy game, some game tokens are in the form of badges of different colour and game power: 'Resistance R (5 badges)', 'Misfortune M (13)' and 'Aid A (7)'. Each player (4) makes a random selection of 3 badges from an opaque bag. What is the probability of the draw progressing, left to right as 12 are drawn as 'M R M - A M R - A M A - R M M?' There are 3 subgroups in total of 25 and a count the incidence of each type of badge in the whole selection gives 6M 3R and 3A. Applying (11.2b) as above,

$$P = {}^{13}C_6 * {}^5C_3 * {}^7C_3 / {}^{25}C_{12} = 0.115493$$

This is for any order and there are $12!/(6!3!3!) = 18480$ arrangements. Each permutation has equal probability and this is obtained by dividing the overall (combination) probability by this total,

$$P(\textit{'in a particular order'}) = 0.115493 / 18480 = 0.00000624964$$

So, there is a 12% chance for getting the twelve in any order but in a particular arrangement it is considerably less, which is logical. So, a very low probability for this particular order.

The with replacement condition

If objects are replaced, the above methods do not apply. All choices revert to the original state probabilities. Each selection is independent and individual probabilities can be multiplied together (the multiplication rule ('AND')). For the classical method, to determine the sample space, we employ the counting principle where n objects are sampled one at a time *with replacement* $= n^r$, where r is the number of selections (Book One Section 7.1.1). Consequently, for the 5 bean selection from a jar, 'taking 1 out then putting it back' over 3 selections can be done in $5^3 = 125$ ways (with order), as there are 5 available each time. For the event, the number of ways of selecting *'1 orange + 2 violet'* can be achieved as:

'orange on first' $= 3$ ways, *'violet on 2nd '* $= 2$ ways *'violet on 3rd '* $= 2$ ways

Remember, beans are replaced and choices are now independent. We want all these to occur together, hence = 3 * 2 * 2 = 12 ways. So,

$$P(\text{'Orange on first, violet on } 2^{nd} \text{ and violet on } 3^{rd'}) = {}^{12}/_{125}$$

This is for one particular order and as seen above this particular selection can be done in 3 different ways, so the final calculation is:

$$P = 3 * (3*2*2)/125 = {}^{36}/_{125}$$

We can also use the individual probabilities of three single selections *with replacement*. The orange beans have a single choice probability of $^3/_5$ and the violets $^2/_5$.

Three events occur in sequence, '*draw an orange AND draw a violet AND draw another violet*'. These are independent, thus:

$$P = {}^3/_5 * {}^2/_5 * {}^2/_5 = {}^{12}/_{125}$$

This is for a particular order and the three beans have 3 (= 3!/2!) different arrangements, so for any order,

$$P = 3 * {}^{12}/_{125} = {}^{36}/_{125}$$

Multiple selections *with replacement*, in any order, can also be modelled by the **multinomial distribution** (Book One Section 11.2) and for the above problem,

$$P = \text{MULTINOMIAL}(1,2) * ((3/5)\text{^}1) * ((2/5)\text{^}2) = {}^{36}/_{125} = 0.288000 \ (29\%)$$

For the 3 from 10 selection,

$$P(\text{'1 of each'}) = \text{MULTINOMIAL}(1,1,1) * (1/10\text{^}1 * 2/10\text{^}1 *7/10\text{^}1))$$
$$= 0.0840000 = {}^{84}/_{1000} \ (8\%)$$

As seen above for single selections, provided we are not specifically concerned with the identity of the other objects in the above and 'with replacement' is used, it is possible to apply the binomial probability function. This can be extended for the multiple case and the probability of getting 0, 1, 2, or 3 of orange on a selection of 1 to 3 can be ascertained. We need to know the probability of success on single selection - in this case it is = 3/5 so,

P('*0 orange on 1*') = BINOM.DIST(0,1,3/5,0) = 0.4,

and the others as P('*1orange on 1*') = 0.6, P('*2 violet on 3*') = 0.288 (both as above), P('*0 orange on 2*') = 0.16 and P('*0 orange on 3*') = 0.06. As seen, the probability of getting zero of orange decreases as more objects are selected.

The multinomial method described above can also be used for dice probability (Section 18.3.3 & Book One formula (11.1b)).

Expected value for multiple selection

This can be calculated using the appropriate parameter of the hypergeometric distribution. For multiple object selections with a main group and one subgroup, the formula used above for a single object applies, with n = the number of selections. With the bean example with 3 selections, the random variable (number of orange appearing) can take the values 1, 2, or 3 (0 orange is not possible) and:

$E(X)$ 'no. orange from 3 orange +2 violet on select 3'

$$= n * M / N = 3 * \,^3/_5 = 1.8$$

Therefore, to the nearest whole number, the number of orange beans to appear in a selection of 3 would average out at 2.

For more than one subgroup, there is the complication of multiple occurrences. Similar formulae apply for each case. Each object is handled individually according to its probability, using the expected value for the multi-hypergeometric:

$E(X_i)$ multi-hypergeo $= n * M_i / N$

For the 10 balls on a selection of 3, expectations would be,

$E(X_i)$ 'no. red on draw 3' $= 3 * \,^1/_{10} = 0.3 \approx$ once every 2

$E(X_i)$ 'no. white' $= 3 * {}^2/_{10} = 0.6 \approx 1$

$E(X_i)$ 'no. blue' $= 3 * {}^7/_{10} = 2.1 \approx 2$

Therefore, for a choice of 3, on average, red sometimes does not appear (once every two passes of the selection of three), white once and blue twice.

For 'with replacment', the multinomial expected value applies in a similar manner with $E(X_i)_{multinomial} = n*p_i$. Thus, in the example, expectations are the same as above, e.g. for i = red, $E(X) = 3 * {}^1/_{10}$, but the probability for a given case will differ between the two circumstances. Looking at the results for each condition from above,

P('without, 1 of each on 3, hypergeometric')

$= {}^{14}/_{120} = 0.116667$ (12%) any order,

cf.

P('with, 1 of each on 3 multinomial') $= 6 * {}^{14}/_{1000} = 0.084000$ (8%)

Thus, expectations are the same but the probability differs.

The expected number of selections to get a particular target object can be calculated given the probability for the conditions of the event. For *with replacement*, the methods described for coins, Section 16.5, based on the geometric and negative binomial parameters apply. Thus for the coloured balls - how many draws with replacement to get a particular colour?

$P = {}^1/_{10}$ and $E(X)$ 'red' $= 1/p = 1/({}^1/_{10}) = 10$

Values for white and blue are 5 $(1/({}^2/_{10}))$ and 1.43 $(1/({}^7/_{10}))$, respectively. On average, a single selection may be enough to get a blue ball, but white and red require many more, reflecting their incidence in the mix.

Calculations for the 'without replacement' state are more involved and are explained in Section 22.3 for card selections. We can use the formula (22.1) given there, to calculate the probabilities, then sum the product of these and the selection stage to get the expected value.

Thus, for blue ball, N = 10, sg = 7, n= 1,2,3 (by 4th selection, P=1)

P($'blue\ ball\ by\ 3rd\ selection'$)

$= 1*7 /(10 -(3-1)) * PERMUT(10-7, 3-1) /PERMUT(10,3-1)$

$= {}^{7}/_{8} * {}^{6}/_{90} = 0.0583333$

Values for the first and 2nd selection are 0.7, 0.233333, respectively, hence,

$E(X) = 1* 0.7 + 1 + 2* 0.233333 + 3* 0.058333 = 1.34$

Just less than when the balls are returned for blue, but a more marked reduction is found on the same calculation for white (3.266667) and red (4.5).

For the expectation of a multiple selection, we can calculate how many draws on average to get a particular set by using the geometric expected value 1/p as above, provided the experiment is repeated over and over. Thus, for the '1 +1 +1' match, we have the probabilities above and,

$'With\ replacement'$ = 1/ 0.084 = 11.904762

$'Without\ replacement'$ = 1/ 0.116667= 8.571404,

giving 12 and 9, respectively.

17.5 'Beans in a jar' games - lottery, bingo and related games

A very popular 'round object' selection process occurs in national lotteries, bingo and some other games and competitions. For some, balls are the vehicle for selection, but others use tickets, entry forms, phone messages, etc. This time, the balls or whatever objects are used, are labelled with numbers or symbols. Many of these games are analogous to the 'beans in a jar' model above and to card selection in terms of probability, with the critical condition of *without replacement* (although 'with replacement' games do occur). Order of selection does not matter for lotteries and most other types, and although bingo draws are announced as they progress, it is the combination of numbers that counts. A key feature of these games is that the objects are distinct and are usually marked with numbers. Players pay a fee for one or more cards, tickets or entries in the hope of achieving a match with the winning selections.

17.5.1 Lotteries

Lottery games have an initial bank of numbers, which are all distinct. These are imprinted on the balls, chits, tickets, etc. These are randomised and a smaller group is selected randomly *without replacement*, by the organizer (manually or mechanically or by computer in video versions). As each number is selected, the sample space, made up by the remaining numbers, gets smaller. Selection of a subset of numbers is governed by this. Probabilities are multiplied but as each event is not independent, subsequent probabilities are dependent. Players choose their own subset independently of the organised draw, in the hope of a match.

For example, in a very simple lottery game, with a bank of 10 numbers labelled as '1' to '10', you mark your ticket with the numbers '4, 7, and 9'. What is the probability of the organizer selecting this particular subset of three as the winning combination? What we are looking for is the chance of a *match* between your numbers and the organizer's:

$$\text{Probability of } 1^{st} \text{ selection} = {}^1/_{10} \quad 2^{nd} = {}^1/_9 \quad 3^{rd} = {}^1/_8$$

$$P(\textit{'all three, in order'}) = {}^{(1*1*1)}/_{(10*9*8)} = {}^1/_{720} = 0.00138889 \ (0.14\%)$$

This is the individual probability method and it generates permutations, i.e. there are 720 permutations of a selection of three numbers from 10. Therefore, as this applies to selection in order, for any order we multiply by the number of permutations of three different numbers = 3!, thus:

$$P(\textit{'all three, any order'}) = {}^{(3!*1*1*1)}/_{(10*9*8)} = {}^1/_{120} = 0.00833333 \ (1\%)$$

Alternatively,

Total ways of selecting (*without replacement*)

$$= {}^{10}C_3 = 120 \text{ combinations (any order)},$$

then specific ways of selecting '4, 7, and 9' *as a combination*

$$= {}^{3}C_3 = 1, \text{ thus,}$$

$$P('all\ three') = {}^{3}C_3 / {}^{10}C_3 = {}^{1}/_{120} = 0.00833333\ (1\%)$$

(or perm. of 3 from $10 = {}^{3}P_3 / {}^{10}P_3 = (3!/0!)/(10!/7!)$

$$= 6 / (10*9*8) = 1/(5*3*8) = {}^{1}/_{120})$$

The multi-hypergeometric can also be applied, formula (11.2b):

$$P = \frac{\binom{1}{1} * \binom{1}{1} * \binom{1}{1} * \binom{1}{0} * \binom{1}{0} \cdots}{\binom{10}{3}} = {}^{1}/_{120}$$

This is the probability of your selection matching that of the organizer – it is very low but there are other chances, as matching less than 3 may win lesser prizes.

The above reasoning and calculation are equivalent to buying one ticket in a '3/10' number lottery and matching the three chosen numbers. Matching of fewer numbers, namely 2- and 1-, are assessed on the basis that each '3-no.' ticket can generate various '2-no.' and '1-no.' tickets according to the total numbers available. A particular original set of 3 numbers generates ${}^{3}C_2 = 3$ sets of two numbers. For example, with your ticket entry as '4-7-9', to win a 2-number prize you need to match '4-7', '7-9' or '4-9' along with a third number that is not called - if any of these pairs are picked you have a '2 number win'. Each of these can be combined with a third number using the remaining numbers – in the current example there are 10 numbers in total; 2 are used for the pair which would leave 8 but 1 of these is the 3rd number on a winning ticket and it cannot be used (otherwise you would have a 3-no. win). The third number is selected from the remaining 7. So, the number of tickets with 2 matching numbers equivalent to one 3-no. ticket:

No. of equivalents $= {}^3C_2 * {}^7C_1 = 3 * 7 = 21$

and P ('2-no. win') $= {}^{21}/_{120} = 0.175$ (18%)

Similarly for a 1-no. match, (your 3–no. ticket generates 3 single numbers, which can be combined with 2 others from the bank,

$$P = {}^3C_1 * {}^7C_2 / {}^{10}C_3 = (3 * 21) / 120 = {}^{63}/_{120} = 0.525 \ (53\%)$$

A much better wager than the 3-no. match although the prize will be less.

This calculation method can be extended to any size of pool and selection, e.g. a 2-number win on 1 ticket in a '3/23' lottery:

$$P = {}^3C_2 * {}^{20}C_1 / {}^{23}C_3 = 0.0338792 \ (3.4\%)$$

National and regional lotteries

The currently popular national and other lotteries come in a variety of versions, according to country and region of origin. The main characteristics of a lottery relate to the total count of numbers available in the pool and the number selected during the draw. A common abbreviation is based on these numbers, e.g. the explanatory example above had 10 available numbers with a selection of 3 and was referred to as a '3/10' lottery. The UK national lottery has 49 numbers and players are allowed to select 6 in the basic form of the game hence this is a '6/49' lottery. The calibrated balls are partially visible through a transparent mixing box; they are randomised by turning this then 6 balls are allowed to tickle through. Applying a combinational formula gives the probability for selection of matching the particular 6-element combination that is generated:

$$P = {}^6C_6 / {}^{49}C_6 = {}^1/_{13983816} = 0.0000000715112 \approx 0.000000072$$

In Excel: $= 1/$ COMBIN(49,6) $= 0.000000072$,

which is the often quoted chance of 'about 1 in 14 million' for winning the jackpot of this lottery. Calculated using individual probabilities for each stage confirms this as,

$$P = {}^1/_{49} * {}^1/_{48} * {}^1/_{47} * {}^1/_{46} * {}^1/_{45} * {}^1/_{44} * 6! = 0.000000072$$

The jackpot prize is the largest, a variable amount but usually several million pounds. With the exception of that for a three number match (£25), variable prizes can be won for four and five matches (around £100 and £1000 respectively, (£2 ticket; new amounts 2013, www.national-lottery.co.uk). Probabilities are calculated using the method described above:

3 no. match, $P = {}^6C_3 * {}^{43}C_3 / {}^{49}C_6 = {}^{246820}/_{13983816}$
$$= 0.0176504 = {}^1/_{57} \text{ (56 to 1)}$$

4 no. match, $P = {}^6C_4 * {}^{43}C_2 / {}^{49}C_6 = {}^{13545}/_{13983816}$
$$= 0.000968612 \text{ (1029 to 1)}$$

5 no. match, $P = {}^6C_5 * {}^{43}C_1 / {}^{49}C_6 = {}^{258}/_{13983816}$
$$= 0.0000184499 \text{ (55490 to 1)}$$

These simplify as fractions for approximate odds expressions. As seen, with a single ticket, a 3-no. match generates almost 250,000 equivalents with a consequent massive increase in the chance of a win. Of course, the win is much smaller and players look to ways to increase the probability for the larger prizes.

Some lotteries have an extra draw and in the UK, a bonus ball is drawn from the same source. If the bonus ball, along with five other numbers on your ticket, match the draw then a bonus prize (around £100000) is awarded. The probability for this event is obtained by looking at the 5-no. match. This can be achieved in 258 ways and the possible matches consist of 6 subsets of 5 numbers from your ticket, each combined with another number from the remaining 43 (hence ${}^{6}C_{5} * {}^{43}C_{1} = 258$ above). Thus, a new 6-no. combination can be formed with any of these original '5 of 6' subsets:

$$P('5\text{-}no. + bonus\ ball') = {}^{6}/_{13983816} = 0.000000429067\ (2,330,635\ \text{to}\ 1)$$

If this is included then the 5-no. match probability is adjusted to,

$$P = {}^{252}/_{13983816} = 0.0000180208\ (55490\ \text{to}\ 1)$$

Many people buy more than 1 ticket and this does increase the chance of a win, thus assuming a different combination on each ticket, probabilities are:

$$P('2\ tickets,\ match\ 6\ numbers') = {}^{2}/_{13983816}$$
$$= 0.000000143022\ (6,991,907\ \text{to}\ 1)$$

$$P('10\ tickets,\ match\ 6\ numbers') = {}^{10}/_{13983816}$$
$$= 0.000000715112\ (1,398,381\ \text{to}\ 1)$$

$$P('100\ tickets,\ match\ 6\ numbers') = {}^{100}/_{13983816}$$
$$= 0.00000715112\ (139,837\ \text{to}\ 1)$$
$$P('7.5\ million\ tickets,\ match\ 6\ numbers') = {}^{7500000}/_{13983816}$$
$$= 0.536334\ (1.2\text{:}1\ \text{in favour})$$

As seen, due to the enormous odds against a win, small numbers of extra tickets are negligible in effect. To get to a 50:50 chance you would need to buy just under 7 million tickets. In addition, the proportional increase gets less - largest boost in going from 1 to 2 where your win probability is doubled, going from 2 to 3 is less (half again) etc. Rather than buy multiple tickets themselves, players can form syndicates to increase chances, with the disadvantage of splitting any win.

The effect of multiple tickets on the probability for the other prizes will increase in a similar manner provided there is no overlap in the combinations contained within the equivalents. This can be ensured by choosing different 6 number combinations on each separate ticket. To win any prize we apply the

'OR' rule and simply add the values for the 3,4,5,6 and bonus ball probabilities:

$$P('any\ prize') = (246820 + 135450 + 252 + 6+1) / 13983816$$
$$= 0.0186376\ (2\%)$$

Thus, odds are 53 to 1 against a win.

Another way of playing this game is to stick to one ticket but play more frequently. Does this give a marked improvement? For example, is 1 ticket per week over one year better than 52 tickets all in one draw? More details are required - any prize? Or just the jackpot?

$$P('get\ jackpot,\ 52\ tickets\ in\ 1\ game') = {}^{52}/\ 13983816$$
$$= 0.0000037185844\ (268,918.5\ to\ 1)$$

We can get the other route by using the addition rule ('OR') rule for winning on 1st week OR second ... all etc. with all the possible combination for win 2, win 3, ... to win all – a long task to get to 'at least 1' jackpot win. Much easier is to calculate the complement: P('*win none of them*'). The probability of 'not winning the jackpot' with 1 ticket is,

$$P = 1\text{- chance of win} = 1 - 0.000000072 = 0.999999929$$

Thus,

$$P('not\ win\ over\ 52\ draws') = (0.999999929)^{\wedge}52 = 0.9999963$$

and the chance of win at least 1 jackpot,

$$P = 1\text{-}\ 0.9999963 = 0.0000037185776\ (268,919.0\ to\ 1)$$

Therefore, there is a tiny, tiny increased chance with 52 draws in 1 game vs. the 1 per week – negligible in practical terms. Similar calculations can be done for the other prizes.

Expected value in lottery

A generally applicable expected value cannot be calculated, as several prizes are variable, depending on ticket sales and number of winners, etc. The probability of losing outright is given by 1 - probability of any win,

$$P = 1 - 0.0186376 = 0.981362$$

Then, if you buy a ticket for £2 and assume a £2 million jackpot, with typical prizes (based on 2013 November levels), a lottery ticket has estimated prize levels (www.national-lottery.co.uk) as £25, £100, £1500, £50,000...

$$ER = -2 * 0.981362 + 2*10^6 * 0.0000000715112$$
$$+ 50000 * 0.000000429067 + 1500 * 0.0000180208$$
$$+ 100 * 0.0009686200 + 25 * 0.0176504 = -1.23$$

(prior to 2013 changes, typical ER = - 0.50 on £1 ticket)

The tickets also include entry into a raffle with a chance of a £20,000 win as,

$$P('Raffle\ win\ per\ ticket') = 50\ /no.\ tickets\ sold$$

Thus, for 10 million sold,

$$P = {}^{50}/_{10} * 10^6 = {}^1/_2 * 10^5 = {}^1/_{200000} = 0.000005\ (1\ in\ 200,000)$$

When this is built into the return, it improves slightly to -1.13 (57%). Sales of course can vary (typical sales 9 million, 6 million in 1 week).

Thus for every £2 spent in this way, about 40% is lost and this increases with lower jackpots. This is the poorest return of all gambling games, but many people play for non-monetary gains (utilities in economic parlance, in the form of participation with other players, etc.).

Several tactics can be employed to help the 'very low' probability difficulty, all probability based:

- Join a syndicate - more entries mean higher probability of a win
- Look at the jackpot each week before play - this fluctuates and higher jackpots would maximise a win
- Use a random rather than a systematic choice for numbers and don't overlap ticket numbers if playing more than one. No overlap means that more numbers are in the mix; random numbers avoid

snags where people use a systematic number choice such as the 'middle column' or 'popular numbers', leading to more chance of having to share with other winners.

- Play in lotteries with higher probabilities (see below)

Several of these tactics have disadvantages, mainly relating to a lowering of the prize value because of sharing, or smaller jackpots from alternative lotteries.

Other lotteries

The recently introduced *UK Health Lottery* game requires players to select 5 numbers from a bank of 50. Prizes are awarded for a full match (5 out of 5), 4 and 3:

$P(\textit{'jackpot'}) = {}^5C_3 * {}^{45}C_2/ {}^{50}C_5 = 0.000000471974$ (2118759 to 1 against)

$P(4 \text{ out of } 5) = {}^5C_4 * {}^{45}C_1/ {}^{50}C_5 = 0.000106194$ ($\approx$ 9416 to 1)

$P(3 \text{ out of } 5) = {}^5C_3 * {}^{45}C_2/ {}^{50}C_5 = 0.00467254$ ($\approx$213 to 1)

These probabilities are considerablely higher than their counterparts for the UK '6/49' game but of course, prizes are smaller. The jackpot ($\approx$ 1 in 2 million chance) amounts to £100,000 or £200,000.

17.5.2 Bingo

With bingo games, there is a variation in that players receive pre-generated cards with selections of subgroups from the bank of numbers. A person, the 'caller', randomly selects each number and declares this verbally or by display, etc. Players tick off the entry on their cards if a match occurs. The first to complete certain sections of a card is a winner. For example, taking our simple instance of a bank of 10 numbers above, a player, Alistair, receives 1 card with three numbers – what is the chance that it has the three numbers that are called first?

A few additional points crop up with this type of game but for now, we will ignore other players and multiple cards. Let us assume that all possible three number combinations have been distributed, individually to players (this amounts to 120). Essentially, the caller will draw three numbers - if Alistair is to win, they must have the same combination as the subset on his card. He has '2, 7 and 5' on his card and the problem becomes exactly the same as above - Alistair has a $^1/_{120}$ (0.008333) chance of his numbers being the first three. If they do not match then one of the other players will have the winning combination if all 120 cards are in the game.

Of course, the bingo game does not usually distribute all the combinations, especially where there are many more numbers in the bank. Players are allowed to have 1 or more cards but some are retained by the banker or the house. The other circumstance is that the caller continues to draw numbers, so in a real game Alistair would not give up after the first three numbers.

Consequently, assuming a non-commercial 'home' game, with only Alistair's card in play - he could win at any point after the first 2 numbers and his chance of doing so is,

3rd number $= 1/^{10}C_3 = {}^1/_{120}$

4th number $= {}^4C_3/^{10}C_3 = {}^4/_{120}$

5th number $= {}^5C_3/^{10}C_3 = {}^{10}/_{120}$

...

10th number $= {}^{10}C_3/^{10}C_3 = {}^{120}/_{120}$

Commercial bingo games

A common form of a commercial UK bingo involves 90 numbers and a card contains 15 numbers. The numbers are arranged on a card containing 27 spaces as 3 rows of 9, so once the numbers are printed there will be some blank spaces. The columns of the card segregate the numbers as 1-10 in column 1, 11-20 in column 2, etc., but the 15 numbers are chosen randomly first before this allocation and this has no effect on the calculation. To win at bingo, players must match the numbers in a line containing one or more rows of 5, (a single line win to all 3 lines, a 'full house'). Multiple plays can be bought (on books of cards or tickets, or on touch pads loaned at the venue) and there are jackpots for getting 'full house' within a certain number of called balls, etc. (www.nationalbingo.co.uk). The number of ways of choosing 15 numbers from 90 to make up a single card is obtained via the combinational formula, ^nC_r, as described above, and it amounts to many, many billions:

$$^{90}C_{15} = 45,795,673,964,460,800 \text{ in Excel}$$

(accurate value $= 45,795,673,964,460,816$)

As might be realised, not all of these are printed out for play use! The game organisers will generate a selection of several 100s, 1000s or more of unique cards (the numbers dependent on whether or not they are for home, party bingo or commercial games).

Calculation of the probability for a match with 1-line requires some reasoning. Five numbers are required and assuming 1 player with 1 card, $^{15}C_5$ gives us the number of ways that the 15 numbers can be formed into groups (combinations) of 5 numbers. To get the probability that the first 5 selected by the caller are on the card, this is divided, via the classical probability formula, with the number of ways to select 5 balls from the 90 in the pool:

$$P(\text{'UK bingo, 5-5 match, 1-line win'}) = {}^{15}C_5 / {}^{90}C_5$$

Using this as the final probability would be incorrect as the 5 numbers could be anywhere on the card – i.e. not necessarily in a row. To get the correct probability we need the latter value and the probability that 5 numbers form a row combination. The 15 numbers selected for a card can be grouped in fives in many ways ($^{15}C_5 = 3003$).

Of these combinations, 1 has a chance of matching the top row and $P = 1/3003$, but there are three rows, so the match with 5 can be achieved in 3 ways – line 1 or line 2 or line 3, hence,

P(*'the 5 called numbers form one of the 3 rows'*) = $3/^{15}C_5$

These two probabilities are linked by applying the AND rule:

$$P = {}^{15}C_5 \, / \, {}^{90}C_5 * 3/{}^{15}C_5 = 0.0000000682605 \; (14{,}649{,}755 \text{ to } 1 \text{ against})$$

Therefore, getting a 1-line win after only 5 numbers have been called, with one card in play, is roughly similar to winning the UK lottery as calculated above. The big difference between the bingo game and lotteries is that more numbers are called. As this happens, the probability improves - all the way until a win is a certainty. The chance of a covering 5 numbers after 6 are called is higher than a 5-5 match, a 5-7 is higher again and so on until by 90 numbers all lines are covered. To get these other probabilities is more complicated and we use the method above for the lottery when matching less than the maximum number for a jackpot. Consequently, for covering 5 when 6 are called,

P(*'5-6 match'*) = ${}^{15}C_5 * {}^{75}C_1 \, /{}^{90}C_6$.

The probability of these being in a line is the same as above and

$$P = {}^{15}C_5 * {}^{75}C_1 \, /{}^{90}C_6 * 3/{}^{15}C_5 = 0.000000361379$$

To this, we need to add the probability of a match of 6 on the card with the 6 called:

P(*'6-6 match'*) = ${}^{15}C_6 \, / \, {}^{90}C_6$

The probability of 5 of these being in a line is obtained by extending the reasoning above to a slightly more involved state. The 15 numbers on the card can form ${}^{15}C_6$ groups of 6. The number of ways that any single row of 5 on the card can be linked with 1 of the other 10 to form 6, is 10 - there are 3 lines, therefore 30 ways in total,

P(*'5 of the 6 are in a row'*) = $30 \, / \, {}^{15}C_6$ and

$$P = {}^{15}C_6 \, /{}^{90}C_6 * 30 \, / \, {}^{15}C_6 = 0.0000000481839$$

The overall probability is the chance of a 1-line cover in a match of 5 with

6 called or a cover in a match of 6 with 6 called:

> P($'5\text{-}5$ match OR $6\text{-}6$ match')
>
> $= 0.000000361379 + 0.0000000481839 = 0.000000409563$

The probability is still very low, but it increases with the number of calls and by 30 calls, there is a chance of about 1 in a 100 (1%) to match any single line and just over 10% at 47 calls. Extending the above style of calculation will show that a single player with 1 card will have an approximate 50:50 chance at call 66.

When the number of calls reaches 10, then a 2-line fill is possible and by 15 a 'full house' can occur. Probabilities for these are worked out in a similar manner but matching 2 rows after 10 numbers and getting full house (all 15 numbers after 15 calls) are even more remote. This latter probability is simpler to calculate, in that we do not need to consider rows - as it comes down to a straight match of the 15 called with the 15 on the card,

> P($'15\text{-}15$ match, full house')
>
> $= {}^{15}C_{15} \ / \ {}^{90}C_{15}$ (1 chance in 45,795,673,964,460,800)

It's not until call 86 that the 'full house' probability nears 50%.

It must be stressed that these probabilities are for 1 player with 1 card! During this time, all the other players will be busy covering their numbers and this of course is the rub! If you were the only player, you could happily await the automatic win when the final number was called. With 2 players what are the relative chances? When 2 players participate, there are now 6 lines available to win and the probability of a win by the 5th call is,

> $6 / \ {}^{90}C_5 = 0.00000014$ 7,324,877 to 1

This large effect (a doubling of the probability) does not continue arithmetically and even by 100 players, $P = 0.00000683$ (146,497 to 1), but with more calls the chances of success shoots up and is just below 50% at 28 calls. These figures are equivalent to a single player using multiple tickets

In addition, chances for a win improve as the number of cards or tickets in the game increases. Cards are sold in books and players can play several books together. In small-scale bingo games at home, low numbers of cards and players are possible but with large bingo halls, there may be many hundreds of players and chances are high for bingo wins. With 100 boards in play, a 1-line fill probability is approx. 56% by call 29. If there are 100 players with 5 cards each, chances of 1-line success exceed 50% on 21 calls. Prize value is usually related to ticket sales but there may be special prizes awarded for events that are not certain to happen, such as a jackpot for a 'full house' call before a low number of calls and other events based on numbers in various positions on the board, etc. Odds for early 'house' claims are very, very remote but at least one has happened on call 23 (*www.telegraph.co.uk*).

Fuller details are given by the 'Wizard of Odds' (Shackleford) and 'Durango Bill' (www.durangobill.com). These sources also provide extensive data for American bingo probabilities. It has 75 numbers and a card that has a 5 x 5 matrix of 24 numbers with an 'unnamed' number represented by an asterisk at the centre. Probability calculations are more complex than those of 90-bingo.

Probability based tactics for Bingo play include:

- Buy more tickets with no overlapping numbers
- Play more tickets than other players
- Play in venues with less players – the call outs are spread over fewer cards

Raffle tickets

Raffle competitions are similar to lottery games, although they use tickets or chits instead of the balls. They are very popular but prize value and number depend on whether they are conducted at local, village setting or at national level. Essentially, numbered tickets are sold. The corresponding stubs are randomised and a random selection is made for 1 or more winning numbers. Application of classical probability is easy as the sample space is made up by the number of tickets, which can be as small or large as convenient. The winning number(s) are decided by the draw, which should be publically viewed. If there is only one winning number, 1 draw is made and,

P('*Your number* = *winning number*') = 1/ no. of tickets

For example, a village fete sells 322 tickets. The stubs are separated and put into a large bowl, then randomised. At the festivities, the bowl is place in front of the merry-makers and the organiser reaches in and draws 1 ticket for the grand prize. If you bought 1 ticket, your chance of winning is,

$P = ^1/_{322} = 0.00310559$ (0.3%, 321 to1)

Alternatively, the hypergeometric function can be us as there is one subgroup of interest:

P = HYPGEOM.DIST(1,1,1,322,0) = 0.00310559

If you bought more tickets, which is usual, you would have more chance, as you have alternative ways of winning, i.e. the 'OR' operative can be applied:

With 2 tickets you can win with,

1st ticket OR the 2nd (but not both) = 2 ways,

therefore,

$P = ^2/_{322} = 0.00621118$ (160 to 1) = HYPGEOM.DIST(1,2,1,322,0)

Similarly, with 5 tickets, P = 0.0155280 (63 to 1) and with 10 tickets, P = 0.0310559 (31 to 1).

If you bought all tickets your chance of winning would be $^{332}/_{322} = 1$, i.e. you would be certain to win, although, out of pocket, unless the prize exceeded the intake! If there are several prizes then there is more chance of getting a single prize or more. Therefore, with 1st, 2nd and 3rd prizes offered, the chance of winning all three with 3 tickets is, via individual probability,

$$P = {}^1/_{322} * {}^1/_{321} * {}^1/_{320} = 0.0000000302355$$

This extremely low value is for a particular order, i.e. you win 1st prize on your 1st ticket, etc. For any order, the number of arrangements is calculated as 3! and,

$$P = 0.0000000302 * 6 = 0.000000181401 \ (\approx 5.5 \ \text{million to 1})$$

The Excel function simplifies this calculation as,

$$P = \text{HYPGEOM.DIST}(3,3,3,322,0)$$

(also by combinational formulae as above $= {}^3C_3 / {}^{322}C_3$)

You would obviously need to buy more tickets for this objective!

The probability of getting a single prize can be ascertained by:

Winning 1st prize OR Winning 2nd prize OR Winning 3rd prize,

With 1 ticket to win 1 of 3:

$$P = {}^1/_{322} + {}^1/_{322} + {}^1/_{322} = 0.00931677 \ 160 \ \text{to 1}$$
$$= \text{HYPGEOM.DIST}(1,1,3,322,0)$$

and, so on. To break 50% chance you would have to buy 67 tickets for at least 1 prize. This and other example calculations are found in Problems & Exercises.

An important difference in raffle ticket draws relates to the efficiency of the randomisation process and there may be more objects to mix compared with those in lotteries. If paper or card tickets are drawn in flat, loose-leaf state, extra care should be taken to ensure randomisation. Rolling each into ball takes more time but makes it easier to generate a random mix. Phone-in competitions, mail shots, etc. are essentially similar to raffles for probability, except the entries are randomised and drawn by computer controlled procedures or random selection of envelopes from bags and bins. Timing of the draw may be critical in some of these competitions.

17.5.3 Scratch cards

These come in an enormous variety of types. They can be based on any of the games mentioned in this book including lottery, bingo, sport, card and board games, etc. Alternatively, they can be composed of a collection of symbols representing anything desired or monetary amounts. The player does not know the procedure used to allocate symbols to the cards but as 'odds of winning' are given, it must be structured such that a certain proportion of winning cards are produced. All symbols or icons are covered on the unused card. The player buys one or more cards and is required to scratch off one or more of the covering coats. This is classic *'beans in the jar selection without replacment'* and with simple forms, probabilities are calculated as above, treating the symbols as subgroups of coloured balls. For example, there could be 20 different symbols allocated to 9 slots in numbers of 0 to 3 on the card, such as,

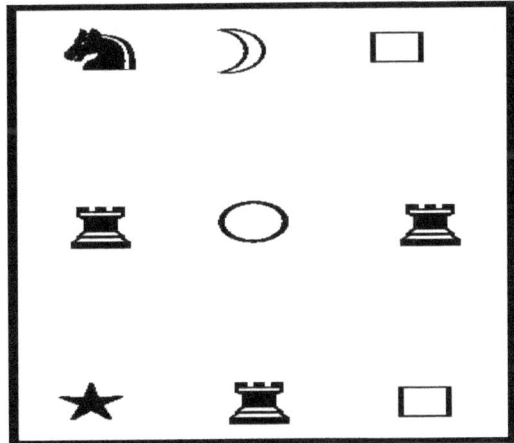

This card has three black rooks, 2 white squares and singles of black star, white circle and white moon and black knight. The card instructs that you should uncover 3 symbols. Three of a kind and two of kind win. What are the probabilities? The formula (9.3c) or the HYPGEOM.DIST( ) function apply. The denominator (sample space) for a choice of three is $^9C_3$ and one event deals with one symbol, namely the rook. Thus, to get 3 rooks,

$$P = {}^3C_3 / {}^9C_3 = 0.0119048 \, (1\%) = \text{HYPGEOM.DIST}(3,3,3,9,0)$$

To get *'any 2 oak'*, formula (11.2b) of the general hypergeometric probability expression is used. The compound event is attained by *'2 rooks and an odd symbol'* OR by *'2 squares plus a single'*, the separate events formed in a union by the addition rule ('OR'),

$$P = (({}^3C_2 * {}^6C_1) + ({}^2C_2 * {}^7C_1)) / {}^9C_3 = 0.297619 \, (30\%)$$

To end up with 'nothing', the player must get 3 different symbols,

$$P = 1 - (P(\text{get } 3) + P(\text{get } 2)) / {}^9C_3 = 0.690476 \, (70\%)$$
(interested readers can attempt to confirm this result by other methods)

If you pay £1 for a card and assuming £5 and £1 prizes for the example events, expected return (with a little approximation) is given as,

$$ER = -£1 * 0.7 + £5 * .01 + £1 * 0.3 = £(-0.7 + 0.05 + 0.3) = -£0.35$$

The positive part of the gain is +0.35, every £1 spent you lose £0.35. Cards like this one have a chance of winning, others in the same set may not. If the proportion of winners in the set is known, the gain can be calculated from the data, e.g. say 25% are winners (5%, £5 and 20%, £1), then,

$$ER = -1 * 0.75 + 5 * 0.05 + 1 * 0.2 = -0.75 + 0.25 + .02 = -0.3$$

Cards may also have a jackpot prize that appears once, which would boost the gain slightly. If this is claimed just after the cards reach the retailers, non-jackpot cards are still on sale, thus it pays to monitor this (www.national-lottery.co.uk/games/scratchcards/prizes).

17.6 Wheels and spins

17.6.1 Roulette

This game has a relatively simple appearance in terms of probability. A rotating wheel has 37 (European) or 38 (American) pockets to receive a single, randomly tossed or 'spun' ball. The individual probabilities for each pocket are $^1/_{37}$ and $^1/_{38}$, respectively. There is a less well-known game, similar to roulette, named after the ball itself: *la boule*. Instead of a spinning wheel, it has a fixed bowl into which the ball is tossed and outcomes are based on nine numbers (1-9).

For *roulette*, Table 17.2 gives the count and probability of the main types of event in play. The pockets comprise 18 each of black and red, alternate numbers of which are odd and even, and one (European) or two (American) green slots. Although primarily a casino game, home versions are available in game compendiums and online in video form, plus it is possible to simulate the wheel using other randomizing devices such as a cardboard spinner, cards or balls labelled with the numbers or by computer simulation (Book One Fig. 14.1).

Table 17.2 Some roulette (European) events and probabilities*

Event	Ways to win	P	Odds against	Casino pays
Straight Up bet (single number)	1	$^1/_{37}$ (0.027)	36:1 36 to 1	35:1
Red / Black / Even / Odd bets	18	$^{18}/_{37}$ (0.486)	19:18 1.06 to 1	1:1
Grouped numbers:				
Street bet (3 nos.)	3	$^3/_{37}$ (0.081)	34:3 11.3 to 1	11:1
Column bet (12 nos.)	12	$^{12}/_{37}$ (0.324)	25:12 2.1 to 1	2:1
Split bet (2 nos.)	2	$^2/_{37}$ (0.054)	35:2 17.5 to 1	17:1
Square bet (4 nos.)	4	$^4/_{37}$ (0.108)	33:4 8.25 to1	8:1
Six line bet (6nos.)	6	$^6/_{37}$ (0.162)	31:6 5.2 to 1	5:1

* ignores re-spins on '0'

To determine probabilities, the classical method is used, by simple counting of qualifying outcomes expressed over the total possible (37 or 38). Thus, there are 18 red numbers on the wheel and P('1 spin = red') = $^{18}/_{37}$ (European). The game sheet in home play or the table in casinos is marked out with the possibilities. Casinos express the odds as 'against' and they take a small contribution for the 'house' so actual pay outs on a win are less than the true odds. The equivalents for a 38-pocket wheel are fractionally less favourable, but can be quickly calculated using spreadsheet functions. Based on the table, bets with the most chance of a win are obviously those near 'evens' where the ways to win are almost the same as the ways to lose. However, whatever the bet the expected return is the same, e.g. with a £1 bet on 'black':

European roulette,

$$ER = - £1 * {}^{19}/_{37} + £1 * {}^{18}/_{37} = -£0.027 \text{ (US} = -\$.053)$$

Casinos give a 1:1 pay out, although true odds are slightly more. For a single number:

European roulette,

$$ER = - £1 * {}^{36}/_{37} + £35 * {}^{1}/_{37} = -£0.027 \text{ (US} -\$.053)$$

Other block number bets are simple to calculate and although ones that are more complex may look interesting, they do not improve the odds. Each part pays off as per the individual bet:

'A square + a single no. within it', (£1 on each),

e.g. '25,26,28,29' with an extra bet on '29'

$$ER = - £1 * {}^{36}/_{37} - £1 * {}^{33}/_{37} + £35 * {}^{1}/_{37} + £8 * {}^{4}/_{37}$$
$$= - {}^{69}/_{37} + {}^{67}/_{37} = -2/37 = -0.054054 \text{ (- £0.027 per £1)}$$

'6 nos. + split exc. 6', (£1 on each),

e.g. '7, 8, 9, 10,11, 12', and a split bet on '35,36'

$$ER = - £1 * {}^{31}/_{37} - £1 * {}^{35}/_{37} + 5 * {}^{6}/_{37} + 17 * {}^{2}/_{37}$$

$$= - \,{}^{66}/_{37} + {}^{64}/_{37} = -2/37 = -0.054054 \, (-\, £0.027 \text{ per } £1)$$

Thus, playing casino roulette ensures a loss in the long term. Nevertheless, the game enjoys great popularity and many players have attempted to devise 'systems' to beat the odds. Some early more successful schemes were based on small but exploitable bias in the wheel. Currently, casinos can subvert this practice by regular servicing and replacment of the wheel.

We can use various Excel functions to get 'multiple spin' roulette probabilities:

P('*a certain no will appear at least once in 37 spins*')

$$= 1\text{-}(\text{BINOM.DIST}(1\text{-}1,37,1/37,1)) = 0.637149 \, (64\%)$$

P('*3 spins, get green, black and red (any order*')'

$$= {}^{1}/_{37} * {}^{18}/_{37} * {}^{18}/_{37} * \text{MULTINOMIAL}(1,1,1) = 0.0383788 \, (4\%)$$

P('*getting 1 in a group of 12 numbers by the third spin*')

$$= \text{NEGBINOM.DIST}(2,1,12/37,0) = 0.148066 \, (15\%)$$

Just like other devices, runs or streaks (Section 15.2.3) are possible on the roulette wheel. For a *simple mixed run* of 5 in 8 outcomes, formula (5.3) is used with run length = 5, fails = 3, p = ${}^{18}/_{37}$, q = ${}^{19}/_{37,}$

P('*run of 5 red with 3 non-red with 8 spins*')

$$= (3 + 1) * ({}^{18}/_{37})^5 * ({}^{19}/_{37})^3 = 0.0147594 \, (1\%)$$

Players may also wonder about how long they need to wait before a particular event(s) appears on the wheel. As illustrated in previous sections (Section 17.4) we can use the expected value of geometric (1/p) and negative binomial (k/p) distributions for such questions. For example,

$$E(X) \text{ '1 from a group of 12 numbers' occurs'? } = 1/\,({}^{12}/_{37}) = {}^{37}/_{12} \approx 3$$

So, on average, it would take 3 spins before '1 in a 12 group' appears. For more than one event,

E(X) '3 odd numbers together in 3 spins' = $3/\left(\left(^{18}/_{37}\right)^3\right) \approx 26$

This is a 'run of 3 odd in 3' and 26 passes (repeated trials of 3 spins each) would go by before the run outcome appeared, on average.

As a gambling example, let's say Raoul (Section 14.3.1 Book One) is back at the roulette table. He has money to spend and wonders about probabilities. He thinks he has a system and decides to bet 37 times on one number, on the basis that it should come up at least once; thereby he will be at least even and possibly ahead. He has a wad of £370 and places £10 on no. 31 and lets it run, replacing if he loses, taking any profit. What is the probability that he will be ahead after 37 plays?

Each loss takes £10 and each win gives £350 back – if he loses 36 times, his money stock would be down to £10 – if he then wins on the last play, Raoul would get back to almost his original at £360. Therefore, to be ahead, he needs to win more than once. It doesn't matter which order these events occur in, so the binomial gives,

P('*at least 1 win on the single no. 31, 37 trials*')
= 1-(BINOM.DIST(1-1,37,1/37,1)) = 0.637149 (64%)

Raoul's strategy may sound good to some but there is a 36% chance of losing on all 37 bets - he assumes that '31' <u>will</u> appear (100%) but it may not; additionally the expected return of his wager is negative.

17.6.2 Fruit (slot) machines and other wheel games

These have 3 wheels (reels or dials) with 10 or more (22) pictures, typically of fruit symbols, on each and pay-outs are given for matching of these symbols in various combinations. The original machines were fully mechanical and the wheels were spun by pulling on the handle (arm) and were limited to the physical structure and symbol markings on each wheel. Nowadays, those in casinos, public houses and recreational areas are more commonly electro-mechanical with software input controlling selection of outcomes via a random number generator. The marking symbols do not necessarily correspond to their incidence on the wheel and are controlled by a 'virtual wheel' with more possibilities. For example, the actual physical wheels may have 20 outcomes on each – and the number of combinations = 20 * 20 *20 = 8000, but the virtual wheel has 64 on each, hence 64*64*64 possibilities. This has the effect of making some outcomes of lower probability (en.wikipedia.org/wiki/Slot_machine).

Home computers, game consoles and tablets have' run alone' or online versions that are full video computerised and practically unlimited in the variety of formats. Modern video machines have many more possibilities and combinations of winning image matches, with more than 3 reels, many more symbols, wild symbols, progressive jackpots and the playing of multiple lines.

Taking the simplest three-reel case, and assuming each wheel is independent of the others, probabilities can be multiplies of the individual wheel probabilities, by use of the multiplication rule (the 'AND' operative). These individual probabilities are decided by the incidence of a particular symbol on the wheel. Events that result in a win can range from a single symbol to combinations of two or more. A typical winning event might be '3 of a kind'. For example, if the symbol of an 'cherry' occurs once on each of three wheels each with 27 symbols, then the chances of getting a named 3, e.g. '*3 cherries*' is,

$$P = {}^1/_{27} * {}^1/_{27} * {}^1/_{27} = 0.0000508 \text{ (very rare)}$$

These devices can be viewed as dice for probabilities – this example is equivalent to,

$$P('a \text{ } particular \text{ } 3 \text{ } of \text{ } a \text{ } kind \text{ } on \text{ } 3d27') = ({}^1/_{27})^3$$

Other wins may be possible with 1 symbol and 2 symbol combinations – this time order may need to be considered in the calculation if particular symbols can occur on any wheels – e.g. 1 'silver duck' – this can be found on wheel 1 or 2 or the third so,

$$P(\textit{'1 silver duck'}) = \frac{1}{27} * \frac{26}{27} * \frac{26}{27} \text{ OR } \frac{26}{27} * \frac{1}{27} * \frac{26}{27} \text{ OR } \frac{26}{27} * \frac{26}{27} * \frac{1}{27}$$

$$= 3 * (\frac{1}{27} * \frac{26}{27} * \frac{26}{27}) = 0.103033 \ (10\%)$$

Usually, it is possible to hold two matching symbols and spin the third for another attempt to convert to win. The probability of the third spin is independent, so it will still be $\frac{1}{27}$ but the overall probability of doing 2 spins and 'getting 2 of a kind' then spinning again to get '3 of a particular kind':

E1 '2 of a kind' – this in order and it can occur in $3!/(2!1!) = 3$ orders ('OOx, OxO, xOO'), hence,

$$P = 3 * \frac{1}{27} * \frac{1}{27} * \frac{26}{27} = 0.00396281 (0.4\%)$$

This probability is given and the probability of 'given that we have two of a kind what is the probability of 3 of a kind?' is equivalent to E1 AND E2,

$$P = 3 * \frac{1}{27} * \frac{1}{27} * \frac{26}{27} * \frac{1}{27} = 0.000146771$$

Slightly more probable than doing on one spin but still quite rare.

The chance of getting 'any win' is given by summing the individual win probabilities (the addition rule ('OR')) – e.g. *I can get a win by getting 3 cherries OR 3 gold bars OR 2 strawberries OR 1 silver duck'*, etc. Of course, each of these wins will usually differ in the amount won. In the real playing situation, you do not know the symbol numbers and types and their distribution. Some symbols may occur on one wheel only, etc. It is possible to log data over many plays and use the relative frequency calculation to get estimates (as was done by Shackleford, who has also designed a virtual machine with known incidence of symbols and hence probability can be calculated (wizardofodds.com)).

Additionally, machines usually provide a *pay table* showing winning combinations and the amount paid out, from which probability can be estimated – the bigger the pay out the more unlikely it is to happen, so the jackpot will usually have the lowest probability. The payment percentage average will not usually be given. If the probabilities are unknown then all a player can do is to gather data on the return as above. The real pay amount ER can then be calculated.

Expected return on slot machines is calculated in the usual manner by summing the product of each outcome (or grouped outcome) and its probability. Assume we have the probabilities (true probabilities differ) for a 3-reel machine with 20 symbols on each (8000 total) and with minimum 'To play' cost = 1 unit:

Event	P	Payout table	Return
1 silver duck	1000 / 8000	1	0.125
3 gold bars (jackpot)	1/8000	1000	0.125
3 cherries	100/8000	50	0.625
No win	6899/8000	0	-1
			-0.125

ER = 0 (-1) * probability of no win + 1000 * probability of jackpot + ...

= - 0.125

For every 1 monetary unit spent, you lose $0.125 \equiv 100 - .125 * 100 = 87.5\%$ - the operator returns just over 12%. Typical expected returns for slot a machine are 90 – 95% of the money put in - again the casino always comes out ahead (howstuffworks.com).

The multinomial can calculate these probabilities. Take the example previous to the above, on 3 wheels with 27 symbols,

P('silver duck + 2 blanks/non-silver duck')

$= MULTINOMIAL(1,2) * 1/27 * (26/27)^2 = 0.103033$ (10%)

More wheels can be accommodated, e.g. on a 5 wheel machine, 27 symbols, - mix of '2 cherries(1) and a star(5) and 2 lemons(9) on 1 'pull'',

$$P = MULTINOMIAL(2,1,2)*(1/27)^2*(5/27)^1*(9/27)^2 = 0.000846754$$

Spinners

Another way to generate random numbers is to use a 'roundish' object device known as **spinner**. It is made with a geometrically shaped piece of card or plastic with the edge demarcated into sectors, the number of divisions is limited only by the readability and confusability as they get smaller. A central spindle enables the device to act like a small spinning top that settles on its edge at the particular sector number. Alternatively, a pointer is mounted at the centre and it is spun round instead. Other devices of this sort include the *driedel* (4 faces) and the *teetotum* (4-12 faces) and a related action is 'spinning a bottle'.

The flat circle shape spinner can be marked out with 2 or more sectors, based on angles – thus with 5 sectors there would be 360/5 = 72° angles, to a give '5-sided' spinner in a pentagon shape with all sides being equiprobable. The probabilities are the same as those for single dice and the roulette wheel and 'wheel of fortune' style games. Some dice are available with the spinner structure (e.g. 50-sided). One spinner design emulated 2d6, with a central circle of 1 to 6 in large digits with a circumference of 6 sets of 1 to 6 in smaller digits (wikimedia.org). The spinner has one very useful feature – the sectors can be of differing angles giving outcomes that are not equiprobable, e.g. 1 sector of 90° (p = $^1/_4$) and 6 of 45° (p =$^1/_6$).

Chapter 18
Dice

18.1 Introduction

Dice, like coins, feature prominently in the history of randomising devices and they have been used in probability related activities for thousands of years. As with any random trial or probability experiment, the dice used and indeed all procedures employed are assumed to be free of bias (Book One Section 3.6). The expression 'loaded dice' is well recognized as a feature of games and activities where this is not the case. With loaded dice, a small weight is buried in the die, such that the weight pulls that face down with more force than the others. This is just one way of the many that can be used to manufacture crooked dice. Even with fair dice, there may deviations from theoretical probabilities due to limitations during production. General gaming dice, which are not manufactured to the same high tolerances as more specialised dice, may exhibit this, but unless very poorly made or worn by use, the effects are likely to be low. Some early versions of polyhedral dice suffered from rapid wear effects.

In contrast, *casino dice* are precision made and any bias will be virtually non-existent. They are usually transparent, which would reveal any interior discrepancies and they are replaced regularly during play use. A characteristic of these dice is the sharp 'square-cut' edges and corners, which ensures minimum rolling on a throw. Many gaming dice have rounded edges and corners partly due to a polishing process, used to remove moulding excess. This state, which increases as dice get worn with use, leads to 'excessive' rolling, especially with the more 'spherical' shaped dice (d20 and d30), which could allow more time for negative effects on fairness. There been various observations on internet sources that this reduces randomness (e.g. www.awesomedice.com). Consequently, a range of sharp-edged dice for the general gamer has been manufactured (Gamescience dice, Lou Zocci; en.wikipedia.org), which have overcome any excessive rolling, but which may require removal of a moulding sprue, normally removed in the polishing process. Thorough shaking of dice in a dice cup, then upending them onto a flat surface may overcome the excessive rolling effect but no controlled study has been performed.

Even with such precision dice, invalidity can still enter into a dice experiment if the throw is not 'true' in the sense of being random. There are well-recognized ways of performing non-random dice throws, especially in the dice game of *Craps*. Inadequate shaking of the dice cup in recreational gaming is another less invidious activity. Thus, assuming fair dice and dice that are at least good quality gaming dice, which are relatively unused, along with adequate randomisation, the probabilities will be valid.

Many games use dice in varying numbers. These include classics such as *Ludo* and *Snakes and Ladders* (one die), the famous *Monopoly®* (two dice), *Risk®* (up to five dice), other board games and many strategy and war and tabletop games (up to many tens of dice and polyhedral dice). *Dungeons &Dragons®* ©Wizards of the Coast, Hasbro Inc. (*D&D®*) and other similar game systems use a variety of dice sizes, typified by the twenty-sided one mentioned above. Naturally, there are many games based on dice alone, such as *Poker Dice* and *Yahtzee®* ©Hasbro Inc. The **binomial model** is appropriate for many but not all dice probabilities.

Dice outcome distributions

There are three forms of distribution that come under this heading, depending on the number of dice and the random variable considered. Single dice outcomes conform to a **uniform distribution**. For multiple dice, distribution of one particular face value occurrence, such as '*the number of 6s on throwing 4d6*', etc. can be modelled by the **binomial distribution** and for two or more faces by the **multinomial** (Book One Sections 8.2, 9.2 &

11.2, respectively for the above distributions). These two cases are illustrated below. Another format of multiple dice outcome data are generated by adding of individual die outcomes (the dice sum). This summation produces another random variable, giving a third type of distribution[3] for dice, that of the sums or totals, which is covered in more detail in Chapter 19.

18.1.1 Dice terminology

Terminology for dice was introduced in Book One Chapter 1 (Section 1.4) but fuller details are necessary. The notation consists of an abbreviation in the form of the 'number of dice' + 'd' + 'no. of faces or sides', e.g. 1d6 is a single six-sided die (cube shaped), and 3d8 are three eight-sided dice. The *size* of a die is taken as the number of sides or faces. Eight-sided dice are an example of one of the polyhedral dice some of which are used in *D&D®* and some other role-playing adventure games.

The number of faces in dice ranges from 4 to 100 and include d4, d6, d8, d10, d12, d20, d30 and d100, although the latter is not a 'proper' polyhedral (Mogenson[a]). The **Platonic solids** are the only ones with exact symmetry (4, 6, 8, 12 and 20 faces but not 10). The more these other dice deviate from the true Platonic solids in terms of symmetry, etc. the less fair they are, where 'fairness' means that all faces have the same probability of appearance.

Six-sided dice faces are almost exclusively marked with pips, spots or dots, although digits appear on some. This latter style is used for non-cubic dice, e.g. a d30 has numbers 1,2,3,4 … 28, 29 and 30 imbedded on the faces. More recently, other forms of dice have appeared: d3, d5 and d7 (crystal/log-shaped) and d14, d16, d24 and d34 and d50 (the latter two with a spinning top shape). The usual 'standard' arrangement for dice labelling is consecutive numbering[4]. Dice can be labelled with symbols other than numbers - names,

---

[3] Another dice distribution is given by the <u>difference</u> between two dice (higher minus lower)

[4] Dice can have non-consecutive numbering configurations, e.g. 2,3,3,4,4,5 (average dice) on 1d6, or numbers plus game specific symbols, etc.

phrases, signs of the zodiac, compass directions, months of the year, hit locations, 'paper', 'rock', 'scissors, etc. Many of these are for games and game purposes and the probabilities are calculated in the same manner as those for the numbered versions.

In addition to the ranges of outcomes with individual dice, combining different dice sizes plus embellishments gives a wide range of face values and sums for game play:

$1d4 + 1 = 2$ to 5 $1d7$: use $1d8$ & ignore '8'

$1d3 = 1d6 \div 2$ $1d100 - 51 = 1$ to 49

(ignore results less than 1)

$2d6 = $ sums of 2 to 12 $1d4 + 1d12 = $ sums of 2 to 16

These represent just some of the scope of the possible outcomes of the dice rolling event. Readers new to polyhedral dice may ask why they are needed when larger totals can be attained by summing cubical dice, etc. This is so, and some of the dice groupings appear to be equivalent in that they have the same range of values, e.g. $1d10+2$ and $3d4$ give the same summed total (3-12) and similarly with $2d8$ and $d4 + d12$ (2-16). This is also true, but the *distribution* and *range* of the values are different (Section 19.2 & Book One Section 14.3.1). This results in differing probabilities for some sums or outcome values (see Fig. 18.2), which have a bearing on the nature of the dice system used in games (Section 20.2).

The abbreviations used for dice outcome notation are conveniently composed by reference to ***partitioning of numbers***, which will be explained in most maths texts. Any integer value can be partitioned or 'split-up' into various component parts, as smaller integers, which when summed give the original number, e.g. 4 can be partitioned into 3+1, 2+2, 2+1 +1, and 1+1+1+1. With dice, we are interested in the partition(s) that correspond to the numbers of dice. Thus, in the previous example, partitions for two dice that add up to 4 would be relevant, namely, 3+1 and 2+2. We may or may not be interested in the sum but the partition concept is applicable to any dice outcome. A particular partition has a combination of integers and these can be notated in a variety of ways, e.g. with two dice outcomes, '31' and '22' is one style and '3,1' and '2,2' is another. This latter style is used in this book for the majority of cases.

For any multiple dice outcome, there is a combination of two or more numbers representing the upper face values on a throw. Each combination can occur in its permutations, thus '3' and '1' can occur as '3,1' or '1,3', etc. This distinction is important and the notation indicates the *order* of occurrence, namely a ***sequence***. Thus, '3,1' means that the first die has a '3' and the second has a '1', representing one sequence of outcomes for two dice. Note that in the case of dice (and coins), these are not 'pure' permutations but combinations (as explained in Book One Section 5.3). It may seem contradictory to say that '3,1' and '1,3' are two different dice combinations, when they appear to be one, as they contain the same numbers. This is because each die is distinct, e.g. we could have a red die and a blue die. Thus, a red '3' + a blue '1' differs from a red '1' + a blue '3'. They can be described as containing the same ***configuration*** (see Fig. 5.1, Book One).

The components of the combinations correspond, of course to the ***elements*** of the single die events. Thus, outcomes for 4d6 summing to '5' would consist of 4 elements, one from each die: '1,1,1,2', '1,1,2,1', '1,2,1,1' and '2,1,1,1', i.e. there are four ways of 4d6 summing to '5'. This particular example has identical elements but remember that each subset is a different outcome.

18.1.2 Dice events

As with coins, events for dice can be in several formats (Section 16.2.1). Single dice events comprise occurrence of a particular face value. For the multiple case, there are compound events with face values or sums for two or more dice. Face value formats give rise to the *content, sequence* and *run* events and adding up the face values gives the very important *sum* event. Table 18.1 lists some possibilities.

Table 18.1 Examples of dice probability events

Occurrence	Example ('probability of getting -')
Single die named face value	An '11' on 1d20
Inequality	More than '89' on 1d100
Singe die range	Between '4 and 9' on 1d10
Multiple requirement	An even no. greater than '2' on 1d10
Multiple:	
Named sequence	'14,22,1' on 3d30
Consecutive sequence	'9,10,11' in order on 3d12, '5,4,3,2,1' on 5d14
Named content	A '7, a 3 and a 5' on 3d8
	'Three 9s and a 13' on 4d14
Mixed dice named content	Chance of getting '20' (1d20), '12' (1d12) and '8' (1d8)
Named content consecutive	'1,2,3,4,5,6' on 6d6 any order
Named content inequality	At least one '6' on 4d6
Named + unnamed content	'Three 7s' on 5d8
Named sequence + unnamed	'6,5,x,y' on 4d6
Run sequence	3 consecutive '19s' on 5d24
Specific sum	A sum of '8' on 2d6
Sum inequality	A total of 'at most 6' on 3d8
Multiple requirement sum	An odd sum, '9 or more' on 2d6
Higher probability?	A '4' on 1d6 cf. an '8' on 1d12
	'2-3' cf. '7-8' on 2d6
Expected value	Average for 7d4, 1d12
Expected no. of throws	to get a '6' (1d6)
Expected no. of throws	To get a double (2d8)
A face value pattern in a game	A 'yahtzee' on 5d6
	A critical hit on 1d20
	A 'farkle' on 6d6
	'Two pair' in poker dice
A geometric sum	Convert a 'point' in *Craps*
A successful manoeuvre in a war game	4 hits of '5 or more' on 9d6

In addition to the above events for specific results, there can be others for ranges, above or below particular values or it can be a comparison of probability values. Other measures include the overall average, the *expected value* (Book One Section 8.2.2) and the expected number of throws required before a specific outcome happens.

Beyond these examples, there are many other configurations of outcomes that can be of interest such as getting various sequences or matches with 2 or more dice, etc. For some problems, not all of the dice will have particular requirements, i.e. they can take *unnamed* values (Section 18.3.4).

This gives rise to a variety of configurations in the event outcomes for dice (Fig. 18.1).

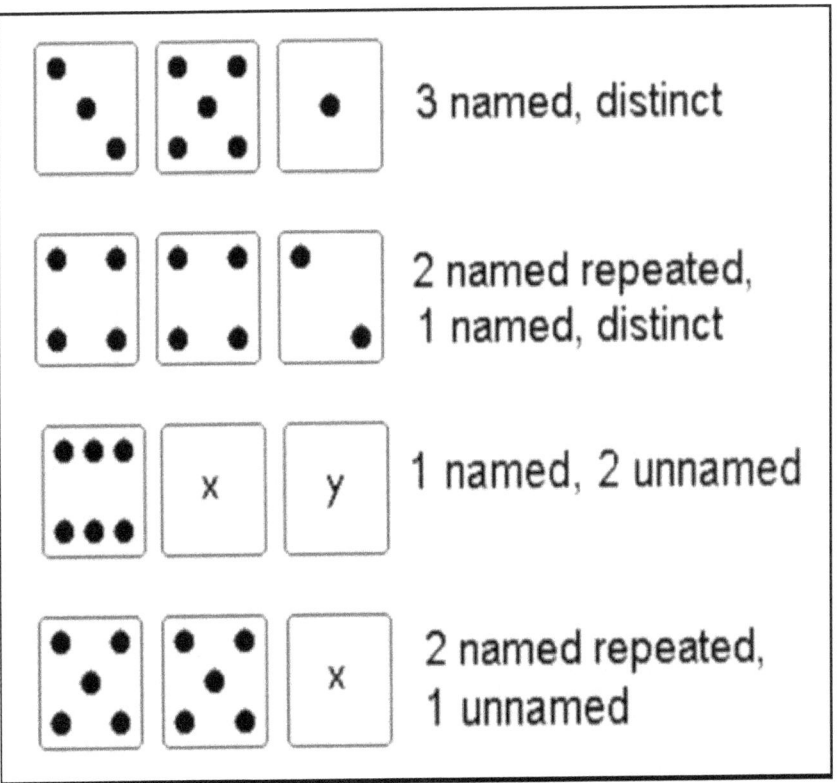

Figure 18.1 Examples of dice (d6) outcome configurations

These events include non-specific outcomes, where one or more dice can take unnamed values as with cards. Many of these patterns are gauged in games such as dice games, war games, and tabletop games. The face values or sums required to 'win' or succeed, depend on the game. In some, particular values achieve success, in others specific ranges are desirable, e.g. *'I need 4 or more to get past the head of that snake – what chance have I got?'* In cases with rolling of multiple dice in some war games getting a '6' signifies a 'hit' e.g. *'10 warrior units attack – 10d6 are rolled – it would be good to get 10 '6s' but is that likely?'*

Note that like coins, where a probability for an 'H event' is the same as that for the T equivalent, dice configurations of the same structure have this property as well. Thus, in Fig. 18.1, any particular three distinct face values on 3d6 will have the same probability as '3,5,1' and any particular pair + 1 distinct value, would have same probability as '4,4,2'. This allows simplification in calculation of probability for general types of pattern with dice, such as 'any 3 of a kind', etc.

18.2 Single dice

Probabilities with single dice are the simplest case and can be calculated with the classical probability formula (3.1) (the number of outcomes in the event divided by the total number; Book One Section 3.2.1). Single dice are used in many games to decide movement of pieces on the play board and to control various actions. Essentially, the number of faces or sides on the die gives the total number of outcomes (the sample space). Determination of the number of outcomes in events also requires relatively simple calculation:

$$Probability_{single\ die,\ any\ face} = {}^{1}/_{no.\ faces} \quad (18.1)$$

This, is of course, the probability mass function of a discrete uniform distribution ($P = 1/(b - a + 1)$) (Book One Section 8.2.3), where 'a' and 'b' demarcate the face values). Thus,

$$P('5 \text{ on a } d6') = {}^1/_6 = 0.166667 \ (17\%)$$

Similarly for $P('11 \text{ on a } d20') = {}^1/_{20} \ (5\%)$, $P('2' \text{ on a } d8') = {}^1/_8 \ (13\%)$ and $P('43' \text{ on } d100\text{-}51') = {}^1/_{49} \ (2\%)$, etc.

The same formula applies to single dice with adjustments, for instance 1d16 + 2 would produce a range of values from '3' to '18'. These would also have a uniform distribution, the only difference being the re-positioning of the upper and lower bounds (Fig. 18.2). Probabilities are $1/(18 - 3 + 1) = {}^1/_{16}$, which is the same as the probability for 1d16. Superimposed on the chart is the distribution for the sum of 3d6. As explained, it has the same range but a quite different set of probabilities.

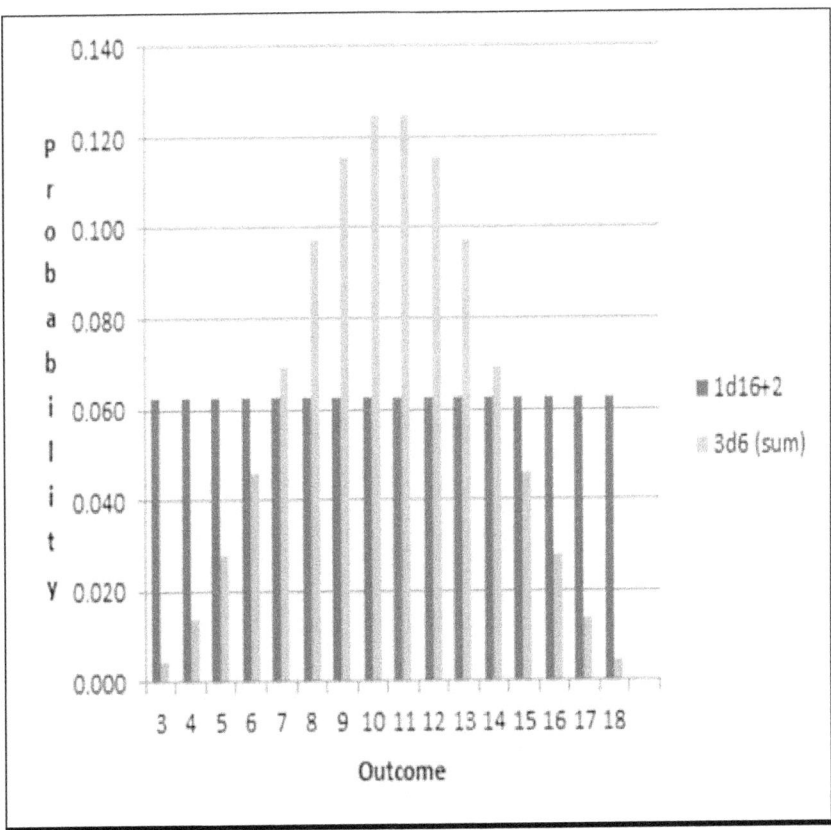

Figure 18.2 Probablity distribution 1d16+2 and 3d6

Although not with a single die, two dice can be used to generate a uniform distribution for a larger range die:

'getting 103 on a 1d120' – this is achieved by allocating 1d12 -1 to the count of tens and 1d10 to the units. Then,

'11 -1' on 1d12-1 ≡ 10 tens = 100 and '3 on 1d10' ≡ 3 units = 3

Hence, P (*'103 on 1d120'*) = $^1/_{12}$ * $^1/_{10}$ = $^1/_{120}$

Similarly, for 1d40,1d60, 1d80, 1d200 and 1d300, etc.

18.2.1 Single dice - ranges and inequalities

The only complication for single dice probabilities arises with ranges and where there are limits imposed as with inequalities. For these calculations, the **addition rule** ('OR'; Book One Section 5.2.2) is applied for the probabilities of the outcomes that qualify for the conditions. In many cases, this is equivalent to a simple count of occurrences. Thus, for *'getting a 3 or more' on a 1d4'* we can say that outcomes '3' and '4' qualify = 2, therefore P = $^2/_4$ = $^1/_2$ = 0.5. Applying the 'OR rule' directly (the individual events are mutually exclusive), then the compound event can occur by attaining *'a 3 on 1d4'* OR *'a 4 on 1d4'* and P = $^1/_4$ + $^1/_4$ = $^1/_2$. Likewise, for *'between 4 and 9 on a d10'* '4,5,6,7,8 & 9' qualify and P = $^6/_{10}$ = $^3/_5$ = 0.6 (60%). With *'more than 89 on 1d100',* ' 90 -100' qualify, and P = $^{11}/_{100}$ = 0.11 (11%). We can use subtraction to get the count, assuming standard numbering and that the range increments by 1 each step, e.g. with '13 or more on 1d24' the count for qualifying outcomes is = 24 -13 +1 = 12 and P = $^{12}/_{24}$. Alternatively, 1 less than the minimum of the range qualifies for 'more than' (at least') events. For events with 'up to' or 'at most' phrases the count is simply the stated value maximum (watch for inclusive or exclusive):

'more than 13 on 1d24' (14 or more) = (24-14+1)/24 = $^{11}/_{24}$

'up to 13 on1d24' = $^{13}/_{24}$

'more than 12 on 1d24' (13 or more) = $^{12}/_{24}$

We should not forget the usefulness of sketches and diagrams of single dice outcomes as illustrated in Chapter 4 (Book One Section 4.3, Fig. 4.4). Single value and range occurrences are readily identified (Fig. 18.3a).

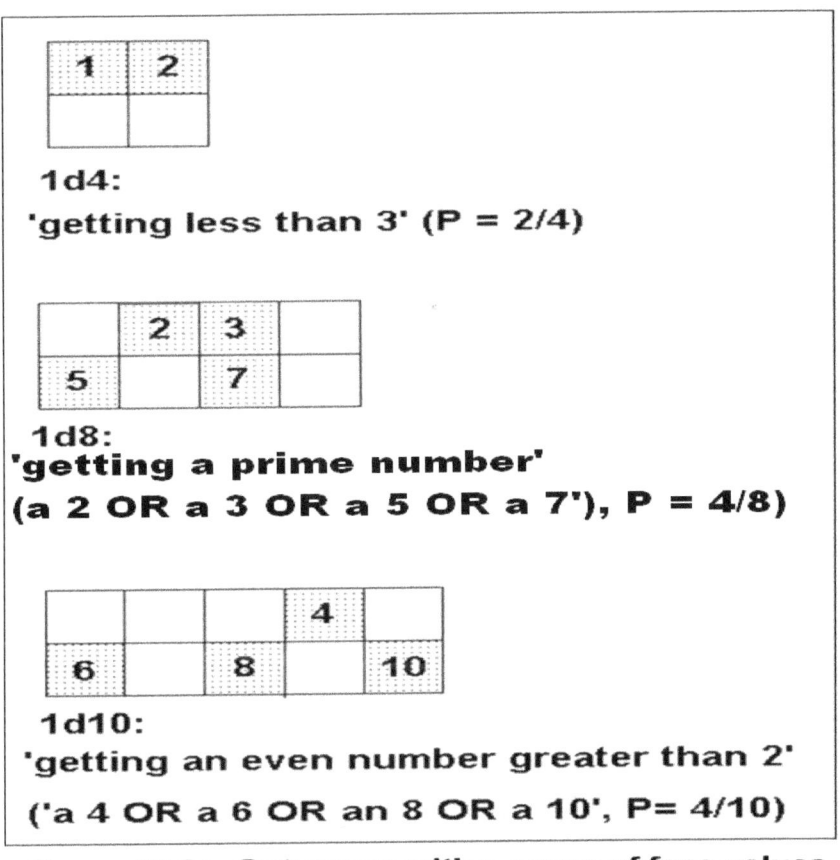

Figure 18.3a Outcomes with a range of face values for single dice

Also, the multiplication rule ('AND') can be applied with single dice where one event depends on another. Figure 18.3b shows this for 1d20, where within the joint event, B depends on A (see below). The sample space is made up by the twenty face values and these are broken up into 'even' and those 'greater than 15'. *D&D*® and some other games use 1d20 for combat results, with a particular face value or more being required for success, based on many factors such as a fighter's experience. We can see that '16 or more' would have probability as event B in the diagram ($^5/_{20}$ = 25%).

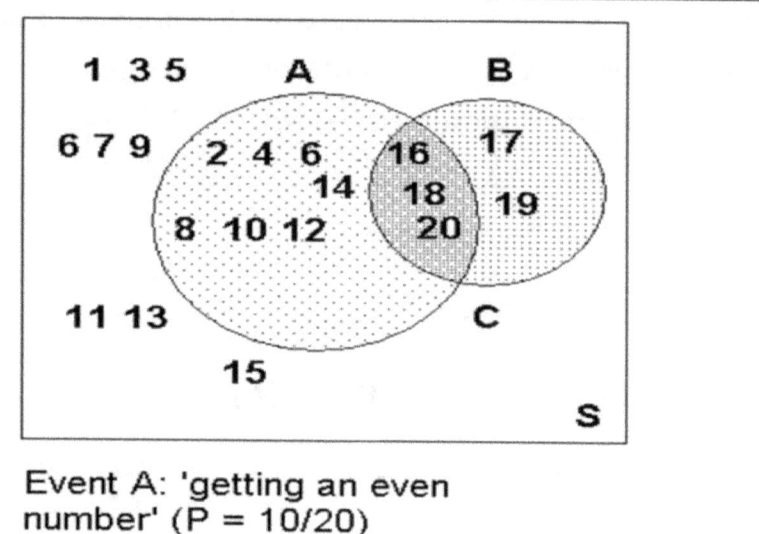

Event A: 'getting an even number' (P = 10/20)

Event B: 'getting a number > 15' (P = 5/20)

Event C: 'getting a no. that is even and >15' (P = 3/20)

Figure 18.3b Venn diagram of events for 1d20

The combined event is represented by the eclipse. The probability is confirmed by the 'AND' rule formula for dependent events (Book One Section 5.2.1):

$P(C) = P(A) * P(B|A)$ thus,

$$P('even \ AND > 15') = P('even') * P('>15 \ given \ even') = {}^{10}/_{20} * {}^{3}/_{10} = {}^{30}/_{200} = {}^{3}/_{20}$$

$P(B|A)$ is given by the common elements of A and B (3) divided by the sample space of A (10).

Expected value and variance for single dice

Single dice outcomes conform to a uniform distribution, so we can use the parameters of this for expected value and variance (Book One Section 8.2.2). The formula (8.3a) for expected value $(E(X)_{uniform} = (b + a) /2)$ can be simplified, as we know that the bounds of the distribution will be decided by the size of the die, plus we know that for the unadorned version the lower bound will always be unity:

$$E(X)_{single\ die} = (s + 1)\ /2$$

In Excel: = (s + 1) /2

Where, s = size of the die *(18.2a)*

The random variable is the face point value, taking values from minimum to maximum, but with a uniform distribution, this is located at the centre. Thus, the expected value for 1d6, given in Book One Section 8.2.2 can be confirmed,

$$E(X)_{1d6} = (6 + 1)\ /2 = 3.5$$

We can easily obtain those for any other size of die, hence,

$$E(X)_{1d8} = (8 + 1)\ /2 = 4.5$$
$$E(X)_{1d30} = (30 + 1)\ /2 = 15.5$$
...., etc.

The formula gives rapid checks on possibilities in games, e.g. in an adventure role-play game of a reptilian creature does 1d12 damage – how many damage points, on average, would it dish out?

$$E(X)_{1d12} = (12 + 1)\ /2 = 6.5 \text{ damage points}$$

If the single die is embellished, then the more general formula needs to be used, thus for 1d6 +3

$$E(X)_{uniform} = (b + a)\ /2, \text{ and } a = 4, b = 9, \text{ then,}$$
$$E(X)_{1d6+3} = (9+4)/2 = 6.5$$
$$\text{and } E(X)_{1d10+3} = (13+4)/2 = 8.5$$

The variance formula for the uniform distribution (8.3b) can also be simplified for single dice:

$$V(X)_{uniform} = (b - a + 2) * (b - a)\ /\ 12$$

For single dice, with s = die size, this becomes,

$$V(X)_{single\ die} = (s - 1 + 2) * (s - 1) / 12 = (s^2-1)/12$$

In Excel: $= (s\wedge2 - 1)/12$

Where, s = size of the die   *(18.2b)*

Thus, for 1d6,

$V(X)_{1d6} = (36-1)/12 = 35/12 = 2.92$ and the standard deviation (SD) = 1.71

For 1d12,

$$V(X)_{1d12} = (12^2-1)/12 = 143/12 = 11.9 \text{ and SD} = 3.3$$

Embellished dice would again require use of the general uniform variance formula, e.g.,

$$V(X)_{1d6+3} = (9-4+2) * (9-4)/12 = 35/12 = 2.92 \text{ and SD} = 1.71$$

Thus, 1d12 and 1d6+3, used for weapon damage in some role-play systems, produce the same average but 1d12 is more variable, which may be used to simulate the differing physical nature of some weapon types.

Another parameter of interest is the **mode** as it identifies the most probable (frequent) outcome. The uniform distribution is multimodal and for single dice all face values are equally likely in the distribution with $P = 1/s$.

Average number of rolls and success by nth trial with single dice

This was illustrated by use of the geometric distribution for 1d6 (Book One Section 10.2) and for coins (Section 16.5). We require the probability value (p) of the trial of interest, e.g. *'a 3 on 1d6'*, then,

$$E(X) \text{ 'no. rolls to get a 3'} = 1/p = 1/ \, ^1/_6 = 6$$

Six rolls on average and this value increases with the size of the die, as with *'how long before I get a 15 on 1d16'*,

$$E(X) = 1/ ^1/_{16} = 16$$

Multiple dice can be treated as a single die to apply this expression, e.g. *'a pair with 2d8'*,

$P = {}^8/_{64} = {}^1/_8$ and $E(X) = 1/{}^1/_8 = 8$ throws

For **success by the nth trial**, the probability function $= (1-p)^{n-1} * p$ is used. Take '*100 on 1d100*' - what is the probability of waiting until 7th roll for this? $p= {}^1/_{100}$, $n = 7$, then,

$P = ({}^{99}/_{100})^6 * {}^1/_{100} = 0.00941480$ (0.94%)

$(P('succeed\ on\ trial\ 1') = 1\%)$

In an adventure game, Doug's character has to climb a wall. He has a 73% chance of success. What is the probability that he succeeds by the third attempt? This means that Doug has 2 fails before he is successful. His fail chance is $1- 73 = 27\%$ and,

$P = ({}^{27}/_{100})^2 * {}^{73}/_{100} = 0.0532170$ (5%)

(His best chance is the first attempt at 73%)

18.3 Multiple dice face value probability

When the event involves more than one die, outcomes can be viewed as counts or sequences of face values or as totals (sums). The latter is dealt with in the next chapter, but for all a count of the *total number of outcomes* is required. This is a simple task to enumerate, but the identity (content) of each outcome is more problematical to elucidate. Diagrams are useful, providing a rapid solution for appropriate cases, but are less effective beyond two dice, where lists and tables are better. Event probabilities can be obtained by the intersection of the independent individual probabilities. Some multiple dice face value problems can be modelled according to the **binomial distribution** (Book One Section 9.2). It cannot provide all answers, as with the sum of dice values and here, other methods, such as more complex **combinatorics** are used.

Multiple dice are found in many game situations, on their own (e.g. *Farkle* and *Craps*), as a tool for movement (e.g. in board games) and for resolution of miscellaneous actions in adventure games, in the form of content, some sequences and sums. Dice sequences occur with games where players are in a circle or line, or in games like *Risk®* where attacker and defender outcomes can be arranged.

18.3.1 Multiple dice – total number of outcomes and their identity

For use of the classical probability equation, we need the sample space, i.e. the total number of outcomes. In the experiment, *'a throw of 3d8'*, how many possible outcomes are there? The basic counting multiplication principle has been used in previous chapters (Book One Sections 4.2.1 & 7.1.1) for this and for multiple dice, it is applied as:

No. of total outcomes$^a_{equally\ sized\ dice}$ = (Size of die(s))$^{no.\ of\ dice}$ = s^n

No. of total outcomes$^a_{any\ sized\ dice}$

= (no. faces, die 1) * (no. faces, die 2) *...

^{a} as sequences

In Excel: = s^n (1 die size) and **= s_1 * s_2 ... (mixed dice***)*

Where, s or s_1, ... = dice size, n = no. of dice   ***(18.3)***

Thus, 4d6 have 6 * 6 * 6 * 6 or 6^4 = 1296 outcomes as sequences and 3d10 have 10^3 = 1000 outcomes.

The formula applies to any number of dice of any size. It gives the count for the number of *sequences,* not *content,* i.e. with 2d6, '5,2' and '2,5' have the same content but they are counted as two unique dice sequences. For different sizes of dice, the latter formula is used, e.g. 1d4 and 1d6 have 4 * 6 = 24 total outcomes; 3d6, 1d8 and 2d12 have 6^3 * 8 * 12^2 = 248832 outcomes. If we want a count of the total number of *content* outcomes, we can use formula (7.5b), namely, $^{n+r-1}C_r$, but note that these will not be equiprobable (see Sections 7.3.2 (Book One) & 18.4).

Multiple dice – event outcomes and identity

The above does not tell us what each of the outcomes in the total count contains in terms of elements. We cannot see the *identity* of the component parts of the sequences or content. For example, a 5d14 experiment has 14^5 outcomes but sequences ending with a '13' are not counted. Similarly, we know that 3d4 have a total of 64 outcomes but if we wish to know how many have a content of 'at least one odd value', we need to enumerate these by other methods. Questions like these and others can be answered by generating a list of all outcomes, typically as sequences.

Manual listing

Preparing a list manually is the simplest way and this will produce all outcomes plus identity. This can be done easily for two dice and tables of this output are legion in probability texts (especially for 2d6!). This text is not exempt, and Table 4.5 in Book One identified the component elements of the outcomes for 2d6 and is referred to many times. A similar table (Table 18.2) lists those for 2d4, where the individual sums are listed first then each subgroup of any two elements that add-up to this sum are filled in.

Table 18.2 Sums, their sequences and probabilities (2d4)

Sum	Sequences	Count	P
2	1+1	1	$^1/_{16}$
3	1+2 2+1	2	$^2/_{16}$
4	1+3 3+1 2+2	3	$^3/_{16}$
5	1+4 4+1 2+3 3+2	4	$^4/_{16}$
6	2+4 4+2 3+3	3	$^3/_{16}$
7	3+4 4+3	2	$^2/_{16}$
8	4+4	1	$^1/_{16}$
	Total =	16	1

Thus, 2d4 has 7 different sums made up from 1 to 4 sequences. Two d6 has

11 sums and more than three times the number of sequences. Diagrams can help with this counting for two dice examples (see Fig. 18.5). Probabilities are obtained by quotient of the individual content or sum count and the total number of sequences, as per the classical formula.

These examples are relatively easy but attempting the same task with more dice gets arduous and errors can crop up due to omissions of instances. Tree diagrams would help avoid such omissions but again they soon get cumbersome with larger sizes and numbers of dice.

All outcome sequences can be generated using more complex spreadsheet functions or by use of macros and programming languages. Essentially, these computer aids generate all possible outcomes and in some versions count the ones of interest.

Dice outcome generation using Excel

With dice, we cannot generate a list of each sequence as easily as was done for coins (Section 16.4.1) but it is possible in Excel and there are numerous ways to achieve this. In the version here (Spreadsheet Table 18.1), a series of counters produce cycles of the outcomes. For example, 3d6 has three cycles, one of 36 (increments by 1 every 36 sequences), another for 6 (increments every 6 times, then resets every 36) and a third for 1(increments by 1 and resets after every 6, 36 times). In the spreadsheet, the sequence number starts in cell C3 (use the fill handle on Fill Series to complete this column) and the following three columns operate the formulas below. Once entered, they can be copied down for the total number of sequences, in this example for 216 rows. The sequences can be built into a text string using:

$$= CONCATENATE(TEXT(D3,0),TEXT(E3,0),TEXT(F3,0))$$

The data table can then be used for searching for partial sequences, runs, and for calculating sums. These can be enumerated by use of the COUNTIF( ) function or Find utility (also possible after transfer to a word processer) with wildcard characters if required. As an example of this, once formed into strings, the STRING column (cell H3 to H218) was searched for sequences consisting of '1,1,x' using COUNTIF(H3:H218,"11?") (= 6) and those beginning with a '6' with COUNTIF(H3:H218,"6*") (= 36).

Spreadsheet Table 18.1

| | | Sequence | 3d6 | | | | |
			d1	d2	d3	SUM	STRING
count	216	(C3) 1	1	1	1	3	111
size	6	2	1	1	2	4	112
c1	36	3	1	1	3	5	113
c2	6	4	1	1	4	6	114
		....	...	...	...	...	
		216	6	6	6	18	666

d1 =1+ INT((C3-1)/36)
d2 = ROUNDUP(C3/6,0)-6*(ROUNDUP(C3/36,0)-1)
d3= C3-6*(ROUNDUP(C3/6,0)-1)

For other dice, the cycle counters can be calculated as, $c1 = s^n / s$, $c2 = c1/s$, $c3 = c2/s$, etc. Thus, 5d6 has 7776 sequences and 4 counters: $c1 = 7776/6 = 1296$, $c2 = 216$, $c3 = 36$, $c4=6$. One limitation is that with more dice the size of a single column style list will exceed the capacity (1,048,576) of Excel – by 8d6 there are 1,679,616 sequences and similarly for 7d8 and 5d12, etc. For such lists, the columns would require to be split and spread across the rows.

Dice sequences can also be identified using a *macro program* to generate all the sequences with some counting. This was done for a small Excel macro (not included in this text) using a series of program loops. Dice size and number are entered and the program generates the list and counts specified events.

Data for sequences of 3d6 were used to prepare a formatted diagram, illustrated in Fig. 18.4. The identity of each component in each sequence is certainly displayed and the diagram can be used for enumeration of event outcomes via the program or by manual counting and search functions.

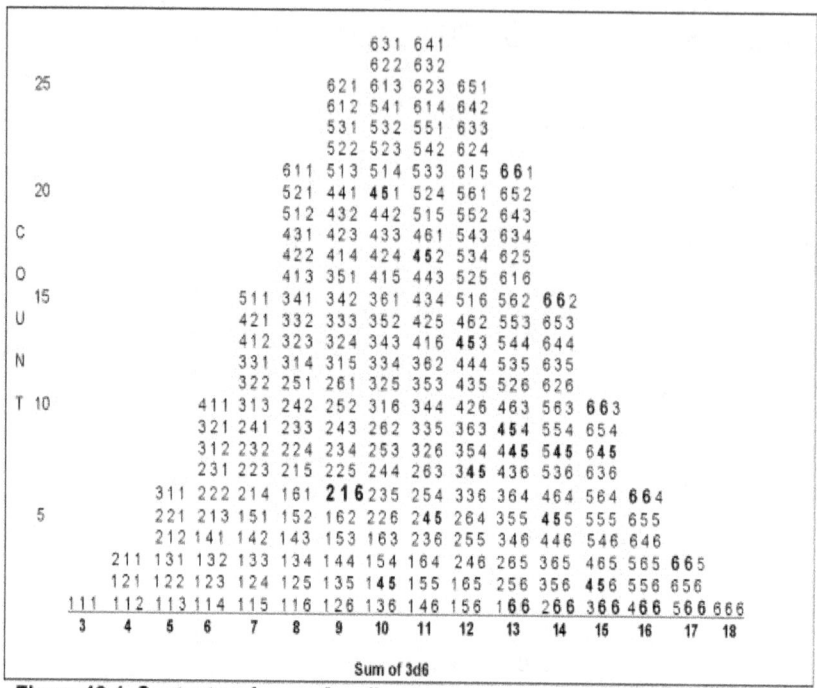

Figure 18.4 Content and sums for all outcome sequences(3d6)

Analysis

As they stand, the outcome lists and diagrams (generated as described above) can be searched without programming for the incidence of any 1, 2, 3 or more component sequence, visually, by printing out and physically marking or by function use. Identified occurrences can be converted to bold font for ease of counting. With Fig. 18.4, any particular 3-part sequence occurs once, e.g. '2,1,6', which coincidentally has a probability of $1/216$. Incidence of 2- and 1-part portions of the three can be located most easily in order, e.g. a sequence with '4,5' together in any sequence – there are 12 sequences that have these components and,

$$P('getting\ a\ sequence\ with\ '4,5'\ in\ order\ on\ 3d6') = {}^{12}/_{216}$$

The manual route can be laborious for some of us, especially so with such diagrams for larger numbers of dice. This difficulty and complexity limits the use of such diagrams for probability problems that require identification of sequence content and counting by computer is much more desirable.

Such data can be used for counting the occurrence of events when no direct formula is available or which involves math that is too complex, e.g. for named and unnamed dice outcomes and run sequences (Section 18.3.4). This type of problem crops up in when calculating probabilities for the combat outcomes in the game *Risk®*. To determine the probability of this event by other means is explained below but most would agree that counting from the diagram, although tiring, does not tax mathematical ability. Fig. 18.4 is used in a similar manner below for a problem involving a dice run, and again in Chapter 19 for sums.

18.3.2 Multiple dice - named sequences

A *dice sequence* result gives the order of appearance of each outcome as the dice are rolled one by one, if such is the intent of the event. For instance, '3,7,12,9' is a sequence obtained on rolling 4d12 where the first die (roll) gives '3', etc. Any dice sequence can be permutated to give all the outcomes with the content of the sequence (Section 18.3.3). Many sequences have the same content, others have a single unique collection of elements. Each named sequence is unique (i.e. they occur once in the sample space). The probability for a sequence is given by the product of individual probabilities (the multiplication rule ('AND')), e.g. '4,2,5' on 3d6 is given by,

$$^1/_6 * {}^1/_6 * {}^1/_6 \, (\,'a \, 4 \, and \, a \, 2 \, and \, a \, 5, \, in \, order')$$

This calculation can be generalised for any number and size of dice as:

*Multiple dice, any named sequence*distinct or repeated:

$$P = 1/s^n \qquad \text{In Excel: = 1/(s\textasciicircum n)}$$

Where, s = dice size, n = no. of dice/ throws
All dice equal size, In order   **(18.5a)**

Thus:

The event '*9,3,7,1,8 on 5d10*' has s =10 and n =5, then

$$P = \frac{1}{10}^5$$

Similarly,

$$P('14,22,1' \text{ on } 3d30') = \frac{1}{30}^3 \qquad P('4,4,4' \text{ on } 3d4') = \frac{1}{4}^3$$

As seen, it does not matter if the sequence contains repeated face values, as the outcome is an ordered occurrence - all values are explicitly identified and the required 'in order' calculation is valid, viz. *'getting '11,19,5,5,5,8' on 6d20'* is a sequence with a some single face values and a triple all in order, etc. Other examples of the phraseology used for such events include:

'*Getting 4,6,3,5 <u>one after the other</u> on 4d6*'

$$P = \frac{1}{6} * \frac{1}{6} * \frac{1}{6} * \frac{1}{6} = \frac{1}{6}^4 = \frac{1}{1296}$$

The latter sequence would constitute a win for the player in position 2 (i.e. die or throw no. 2) in a game where four players roll a single die in turn and the highest roll wins. To get the overall probability for the player all sequences where he/she wins would have to be summed (Section 15.2.1). A similar situation occurs in battles in the game of *Risk®*, with dice outcomes split into two parts, e.g. the first three for the attacker and the last two for defender:

'*Getting '5,6,3,1,2', <u>in a row</u> on 5d6*'

This sequence would constitute one version of an 'attacker win' as there are two higher face values in the first three dice. The first three outcomes are for the attacker's dice and are considered separately (the order applies to the two groups of dice and within each group there can be any order).

The probabilities of the individual sequences above are low but this has no bearing on the interpretation of the results in these examples as they are but one of a number of sequences that would achieve the same result.

Sequences with different sizes of dice

If a collection of dice with differing numbers of faces is used in a trial, then the individual dice size(s) must be used in the calculation:

Multiple mixed dice,

any named sequence$_{distinct\ or\ repeated,\ in\ order:}$

$P = {}^1/_{die\ size\ no.1} * {}^1/_{die\ size\ no.2} * {}^1/_{die\ size\ no.3 ...}$

Dice of same size or mixed sizes

In Excel:

= (1/s_1) * (1/s_2) *(1/s_3) ...

Where, s_1, s_2, etc. are die sizes   **(18.5b)**

Thus, for '*11,15,23,9,7*' on 1d12,1d20,1d30,1d10 and 1d8,

$P = {}^1/_{12} * {}^1/_{20} * {}^1/_{30} * {}^1/_{10} * {}^1/_8 = 0.00000173611$

A varied selection of dice but an extremely rare event (Problems & Exercises).

The event is more clearly stated as '*an 11 on 1d12, then 15 on 1d20, then 23 on 1d30 ....*'.

For a sequence of '*17,7,9,3*' on 2d20, 1d10 and 1d4,

$P = {}^1/_{20}{}^2 * {}^1/_{10} * {}^1/_4 = 0.0000625000$

The outcomes must match the dice size, i.e. the event '*getting the sequence 17,29, 4 on 1d4,1d30, 1d12*' has a probability of 0 as it is not possible to roll a '17' on a four-sided die.

The relatively simple calculations above for named dice sequences apply, in part, to probabilities for dice content, but an additional step is required.

18.3.3 Multiple dice - named content

This refers to one or more face values occurring for two or more dice, e.g. *'getting a 3, a 5 and a 6 with 3d6'* or *'a 14 and a 6 on 2d20'*. Order is *not considered*, thus any sequence that has this content qualifies. These events are commonly referred to as *'dice combinations'* but this description is also used by some for dice sequences and reference as *dice configurations* avoids this confusion, bearing in mind the points above.

Content can include all distinct face values as well as repeated ones: *'getting 'a 9 and two 10s' on 3d14'* describes an outcome content with one repeated value (a double or a pair).

Most games specify content (or sums) regarding the requirement, e.g. *'any die with a 5 or 6 in a throw (of 10d6) wins the action'*, or *'one scoring pattern is a straight (consecutive integers) of 5 numbers'* – the numbers can occur in any order.

The calculation does depend on how the event is stated in this respect. Thus, statements like *'getting a 3 on 1d6 <u>and</u> a 4 on another'* could be calculated as $1/6 * 1/6 = 1/36$, and this would be correct if it is assumed that the '3' was to be before the '4'. If, on the other hand, order is not stipulated then the outcome can occur as '3,4' or '4,3' giving a probability, with *any order* equal to $2/36$, using the 'OR' case with addition of the probabilities.

For the 'any order' circumstance, as an initial step, we can easily calculate the probability for a sequence as above using the individual probabilities, but this requires to be multiplied by the number of orders possible, as calculated by an appropriate permutational formulae:

Multiple dice, any named content:

Distinct: $P = 1/s^n * n!$ In Excel: = (1/s^n) * FACT(n)

Repeated: $P = 1/s^n * n!/a!b!c!...$

In Excel:

= (1/s^n) * FACT(n)/ (FACT(a_1) * FACT(b_1) * FACT(c_1) ....)

or = (1/s^n) * MULTINOMIAL(a_1,b_1,c_1, ...)

Where, s = dice size, n = no. dice

a,b,c, .. or a_1,b_1,c_1 = nos. rep. & dist. occurrences   **(18.6a)**

For named values that are all distinct, $n!$ is used ((7.3a), Book One Section 7.2.1). Thus '3,5,6' can occur in 3! (= 6) arrangements, hence:

$$P('a\ 3,\ a\ 5\ and\ a\ 6\ content\ on\ 3d6') = \frac{1}{6}^3 * 6 = \frac{1}{36} = 0.0277778\ (3\%)$$

'Getting a 3, a 6, a 1, a 2 and a 5, in any order on 5d6':

$$P = \frac{1}{7776} * (5!) = \frac{120}{7776} = 0.0154321\ (1.5\%)$$

'Getting a 7, a 3, and a 5, in any order on 3d8':

$$P = \left(\frac{1}{8}\right)^3 * 6 = 0.0117188\ (1\%)$$

This method can be used when each value is distinct, but if repeats are included then the 'permutations with repeats' formula (7.3b) (Book One Section 7.2.2) must be used:

For 'getting 'two 4s, a 2 and a 1 with 4d6 in any order', $n = 4$, $a = 2$, $b = 1$, $c = 1$, then,

$$P = \frac{1}{6} * \frac{1}{6} * \frac{1}{6} * \frac{1}{6} * 4!/(2!*1!*1!)$$

$$= \frac{1}{1296} * \frac{24}{2} = \frac{12}{1296} = 0.00925926\ (1\%)$$

In a game of *Farkle*, 'two 3s, a 2, a 6, a 4 and a 1 with 6d6 in any order' are rolled:

$n = 6$, $a = 2$, $b = 1$, $c = 1$, $d = 1$, $e = 1$

$$P = \left(\frac{1}{6}\right)^6 * 6!/(2!*1!*1!*1!*1!)$$

$$= \frac{1}{46656} * \frac{720}{2} = \frac{360}{46656} = 0.00771605\ (1\%)$$

This is for a particular scoring roll in the dice; to get a more general *'a 1* with a pair (2,3,5 or 6) and another three distinct values except 5', requires multiplication by 4, $P = {}^{1440}/_{46656} = 3\%$.

As dice size increases, base probabilities in the named content get lower:

$$P(\text{'two 47s and a 29 on 3d50'}) = {}^1/_{50}{}^3 * 3!\,/2! = {}^3/_{50}{}^3 = 0.000216000$$
$$P(\text{'three 23s on 3d30'}) = {}^1/_{30}{}^3 * 3!\,/3! = 0.0000370370$$

Readers may observe that the factorial expressions in the latter examples are equivalent to the **multinomial coefficient** (Book One Section 11.2) which indeed can be applied to these data. Thus, for the first example above, using Excel:

$$P(\text{'two 4s, a 2 and a 1 with 4d6, any order'})$$
$$= (1/6)^4 * \text{MULTINOMIAL}(2,1,1)$$
$$= {}^1/_{1296} * 12 = 0.00925926 \ (1\%)$$

With more repeats - 'getting 'two 7s, a 9, three 11s and a 2 on 7d12 in any order',

$$P = {}^1/_{12}{}^7 * 7!/(2!*3!*1!*1!)$$
$$= (1/12)^7 * \text{MULTINOMIAL}(2,3,1,1) = 0.0000117214$$

'Four Js and a K' in poker dice,

$$P(\text{'J,J,K,J,J', poker dice}) = {}^1/_6{}^5 * 5!/(4!*1)$$
$$= {}^1/_{7776} * 5 = {}^5/_{7776} = 0.000643004 \ (0.1\%)$$

Again, a particular instance of a pattern, 'any four of kind' is much more probable, $(P = 0.000643004 * 6 * 5 = 0.0192901 \ (2\%))$.

Many dice content problems of the above sort can also be solved using the binomial formula, provided that all outcomes are named and consist of repeated instances (Section 18.3.4). For example, the 3d30 instance above represents 3 successes on 3 trials and with the binomial Excel function:

$$P = \text{BINOM.DIST}(3,3,1/30,0) = 0.000037037$$

Similarly, achieving a content combination of *'seven 2s on 7d6'*,

$$P = (^1/_6)^7 * 7! /7! = ^1/_6{}^7 = 0.00000357$$
$$= \text{BINOM.DIST } (7,7,1/6,0) = 0.00000357 \ (^1/_6{}^7)$$

The other examples above would not come into this category, i.e. it is not possible to solve *'getting 'two 4s, a 2 and a 1' with 4d6 in any order'* using the binomial (the binomial application is discussed more fully below).

Outcome content for mixed dice sizes

These are dealt with as above, except that the sizes are detailed in the event. There is a complication in that one or more of the outcomes stipulated may not apply to every die thrown in the trial. Thus, to get a content of *'a 4 and an 8 on 1d20 and 1d8'* would equal $^1/_{20} * ^1/_8 *2! = ^2/_{160}$ but a content of *'a 19 and an 8 on 1d20, 1d8'* would be $^1/_{20} * ^1/_8 = ^1/_{160}$. This is also the correct probability for the event with order, but with different sized dice and explicit allocation to them, order does not matter, i.e. it is not possible to get '19' on d8 so there is no other order possible. The previous event can occur in either order.

Similarly,

'getting a 20, a 12 and an 8, specifically on 1d20, 1d12 and 1d8':

$$^1/_{20} * ^1/_{12} * ^1/_8 = ^1/_{2020}$$

This is the probability for the event in order, as we cannot allocate the '20' to either of the other dice, etc.

Also, *'getting 'a 3, a 6, a 1, a 4, and a 5' in any order on 1d6, 1d8, 1d14, 1d20 and 1d30'*. In this case, outcomes are possible on all the dice and with 5 distinct occurrences,

$$P = ^1/_6 * ^1/_4 * ^1/_8 * ^1/_{20} * ^1/_{30} = ^1/_{115200} * (5!) = ^{120}/_{115200}$$

For more elaborate examples of this sort, where the clash does happen, permutations for order are calculated for the appropriate dice. Thus, for a content of '*a 4, a 5, and a 27'* with 1d6, 1d12 and 1d30 we can see that a '4' and '5' can be attained by any of the three dice, but the outcome '27' applies only to the 1d30. Thus,

$$P = \frac{1}{6} * \frac{1}{12} * 2! * \frac{1}{30}, \text{ in Excel as,}$$

$$= 1/6 * 1/12 * \text{FACT}(2) * 1/30 = 1/1080 = 0.000925926$$

For repeats where all face values are possible on all sizes of dice used,

P(named content)$_{mixed\ dice:}$

= 1/(s_1 * s_2 ...)*MULTINOMIAL(a_1,b_1,...)

Where, s_1, s_2, ... = different sizes of dice

a_1, b_1, ...= count of face values   **(18.6b)**

For example, P(*'content a 3 and two 4s on 2d4,1d6'*),

$$= 1/(4*4*6) * \text{MULTINOMIAL}(2,1) = 1/96 * 3 = 1/32 = 0.0312500\ (3\%)$$

Diagrams and dice face outcomes

Dice sequence and content probability can also be elucidated with diagrams, which provide relatively simple solutions for some problems. This applies mainly to the two-dice situation where two way matrices can be used (Book One Section 4.3). These matrices are very useful and can quickly identify and enumerate complex outcomes by just counting the markers. Simple instances of content and sequence are shown (Fig. 18.5). The diagram includes cases where the outcomes are fully specified and others with some unnamed. Also illustrated are the union and intersection ('OR' and 'AND', respectively) of some probabilities, as well as for a particular sequence format (consecutive numbers) and a display of the outcomes related to a two-player dice game (see Section 15.2.1).

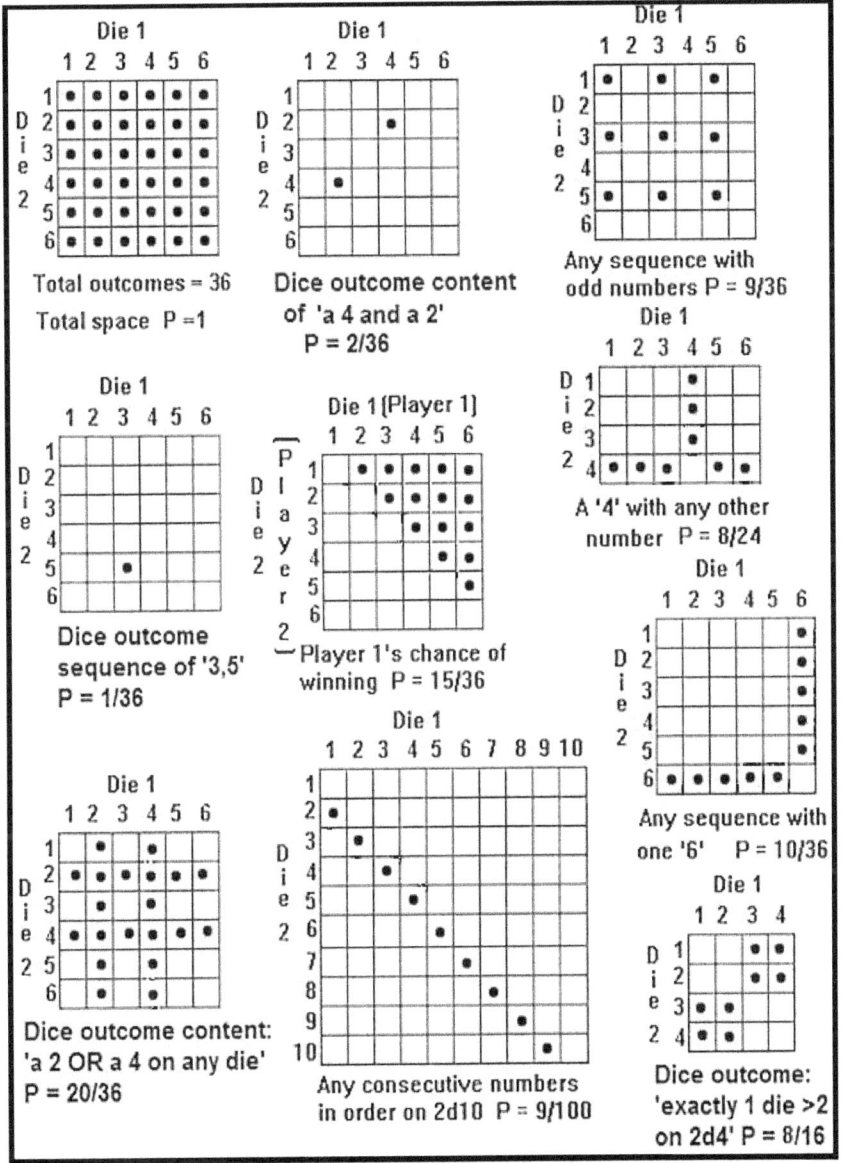

Figure 18.5 Two-way matrices for 2 dice probabilities

One of the more complex examples is where the diagram is extended to elucidate an 'OR' operative application, namely, *'a 2 OR a 4 on any die'* (P = $^{20}/_{36}$). This probability can be confirmed by manual counting from the dice table in Book One (Table 4.5). The problem is quite involved to work out using reasoning and formulae:

A '2' on die 1 AND any on die 2 = $\frac{1}{6} * \frac{6}{6} = \frac{6}{36}$

    OR

A '4' on die 1 AND any on die 2 = $\frac{6}{36}$

    OR

A '2' on die 2 AND any on die 1 = $\frac{6}{36}$

    OR

A '4' on die 2 AND any on die 1 = $\frac{6}{36}$

All these qualify and can be added = $\frac{24}{36}$, but with formula for events that are not mutually exclusive, the 'both' state occurs and this must be subtracted, as it has already been included:

A '2' on die 1 AND a '4' die 2 = $\frac{1}{6} * \frac{1}{6} = \frac{1}{36}$

    OR

A '4' on die 1 AND a '2' die 2 = $\frac{1}{36}$

    OR

A '2' on die 1 AND a '2' die 2 = $\frac{1}{36}$

    OR

A '4' on die 1 AND a '4' die 2 = $\frac{1}{36}$, added = $\frac{4}{36}$,

$$P = \frac{24}{36} - \frac{4}{36} = \frac{20}{36} = 0.555556 \ (56\%)$$

This is a good example of non-mutually exclusive events – both parts require subtraction - counting the markers in the diagram avoids this possible error. Unfortunately, applying this type of diagram to three or more dice is difficult, and leads to complexity in interpretation.

Consecutive number outcome

A special case of a dice sequence (or content) is the *consecutive number outcome* (a 'straight' in some dice games parlance). For the ordered type of example, probabilities are obtained with the general formula for a dice sequence and:

P(particular consecutive numerical sequence) = 1/s^n

In order $\qquad$ **(18.7a)**

(For this formula and the related ones below, s = size of dice, n = no. of dice)

Thus, '1,2,3' in order with 3d6, would be $\frac{1}{6}^3 = \frac{1}{216}$ and for '9,10,11' on 3d12, $P = \frac{1}{12}^3$. When stipulating 'in order', the event would need to say ascending or descending order, e.g. '3,2,1'.

For a straight as *content,* we multiply the sequence probability by the number of arrangements as done above - all face values are distinct, so the number of sequences is always n!:

P(particular consecutive numerical content) = 1/s^n * n!

Any order $\qquad$ **(18.7b)**

Thus for '1,2,3', we can have:

'1,2,3' or '1,3,2' or '2,1,3' or '2,3,1' or '3,1,2' or '3,2,1' and

$$P = \frac{1}{6}^3 * 3! = \frac{6}{216} = \frac{1}{36} = 0.0277778 \ (3\%)$$

'Getting exactly '1,2,3,4,5,6' in <u>any order</u> with 6d6'.

$$P = \frac{1}{6}^6 * 6! = \frac{720}{6}^6 = 0.0154321 \ (\text{approx. 2\% or } \frac{1}{65}, \text{ a straight in } Farkle)$$

Usually in dice games, the dice will be undistinguishable and the order of the straight will be unknown unless the dice are rolled singly. As stated above, this does not matter and content occurrence is sufficient, thus in *Yahtzee®* '2,6,4,5,3' counts as a 'large straight' content. This game allows a 'partial straight' in some rule variations (Section 20.2.1).

A further complication is where 'any sequence' is specified in the event – for 3d6 this would mean '1,2,3', or '2,3,4' or '3,4,5' or '4,5,6'. This broader specification requires an extra calculation:

No. of <u>full</u> straights possible = s - n +1 (n<= s) **(18.7c)**

For 2d6, No. = 6 – 2 + 1 = 5, 5d6 = 6-5+1 = 2, etc.

This calculation applies as long as the number of dice does not exceed the number of faces. When these parameters are equal there is but one consecutive numerical sequence possible. So, this applies to 6d6, 4d4, 12d12, etc. Once this limit is exceeded, a full group of consecutive numbers is not possible, e.g. we cannot get 7 consecutive numbers with 7d6. With different sized dice (less or more faces cf. d6) the possibilities change – e.g. for 3d4 there are 4 - 3 +1 = 2 straights; for 7d20 there are (20– 7) + 1 = 14 straights of length 7 and so on.

Coupon collecting

Multiple distinct numbers in throws of dice simulate the occurrence of distinct objects in boxes and packets, as per the situation for coupons, toys, special items, etc. in packets of food, other retail goods or in starter and booster packs of game cards. Each throw (die) represents an acquisition or purchase of 1 packet that has a chance of containing one of the randomly distributed objects (represented by a particular face value). Thus, if there are 6 items in the collection, the probability of buying 6 and getting a different item each time to complete the set is,

$$P(\text{'buy 6 to get all 6'}) = 1 * \tfrac{5}{6} * \tfrac{4}{6} * \tfrac{3}{6} * \tfrac{2}{6} * \tfrac{1}{6} = 0.0154321$$

This is an example of the **'coupon collecting problem'**. We will recognise the calculation above as that for a dice sequence with 6d6 and as similar to cards with a reducing sample space. Using the dice analogy assumes a large population of objects so that selections make negligible effects on probabilities. Hence the multinomial function can be used to calculate the above (but see below). The result compares closely to that by the multi-hypergeometric in Book One Section 11.3, for 'without replacement' conditions and an assumption of a <u>defined</u> large population. Note: both functions and direct calculations give the correct result when no. items = no. purchases, but require modification when the count of purchases exceeds items due to their 'any order' nature. For example, to get '6 on 7 purchases' means that one item appears twice, any time after the first but not by the last.

This probability cannot be calculated by 'getting 5 different face values and 1 pair on 7d6' because the pair must occur by the 6th purchase. 7d6 sequences with 6 consecutive face value (positions 1 to 6) must be subtracted from the total or a modified calculation used.

Of interest in such problems, is the <u>average no.</u> of packets to buy to get all the items - useful to know before you embark on a collecting hobby! This can be ascertained by calculating the expected value for each action separately, then applying the additivity of E(X)s ((8.1b), Book One). The random variable is the numbers of purchases or acquisitions to get the item. Assuming a large population, we can use the geometric distribution. The expected value for the first purchase is 1 - you are certain to get a starting item,

$$E(X) \text{ 'get first'} = 1*1$$

This leaves 5 to get and with the geometric you can get it in 1 purchase or you can fail:

Succeed on 2nd purchase $= \frac{5}{6}$

    OR

Fail on 2nd purchase $= \frac{1}{6}$

This means you get a match with the one that have already, then you can get one on the next choice or fail again, etc. The average number of purchases to get the second is obtained by the geometric expected value,

$$E(X) \text{ '2nd'} = 1/p = 1/(\tfrac{5}{6}) = \tfrac{6}{5}$$

Likewise for the others and,

$$E(X) \text{ 'all '} = 1 + 1/\tfrac{5}{6} + 1/\tfrac{4}{6}... = 1 + \tfrac{6}{5} + \tfrac{5}{4} + ...$$

Apply this to all and,

$$E(X) \text{ 'get all 6'} = \tfrac{6}{6} * \tfrac{6}{5} * \tfrac{6}{4} ... \tfrac{6}{1} = 14.7 \text{ (15 packs)}$$

This type of calculation is conveniently given by in Excel by the **array** formula {SUM(6/(ROW(1:6)} (enter the expression then press **Shift-Ctrl-Enter** together). The first '6' represents the total number of items in the full set and the ROW(1:6) results in the expression above with division by 1 to 6.

So, for the 'bubble gum cards' of past times, with 100 pictures of different aircraft, baseball players, etc.,

$$E(X) = \{SUM(100/(ROW(1:100)\} = 519 \text{ cards}$$

Sadly, I and many fellow collectors of such sets never completed our collections but had many repetitions for swapping. Collections with less in the set are more attainable,

Game card purchases to get all rare card (30) = 119.85 (120 cards)

The probabilities for getting several or all of the required type fall as their number increases – buying 100 and getting all is equivalent to rolling 100d100 and getting 100 distinct numbers - an extremely low chance (= $100! / 100^{100}$).

18.3.4 Multiple dice - named and unnamed outcomes

For the above examples, all face values are *named*, but it is possible to include unspecified ones, i.e. *unnamed* outcomes, as shown in Fig. 18.1. For example, *'a 7, a 3 and two other numbers on 4d8'* are named outcome contents for two dice of four and the remaining two can take any of the possible face values, except the named ones (a common assumption). These combined events take several configurations where the named outcomes can be distinct or repeated, along with other unnamed occurrences:

'three 7s' on 5d8' = '7,7,7,x,y'

'nine 6s on 10d6' = '6,6,6,6,6,6,6,6,6,x'

'a '5' and a '9' on 7d10', one instance' = 'v,9,w,x,5,y,z'

'two 4s and two 2s on 6d6' = '4,4,2,2,x,y'

Such problems can be solved by a variety of methods, the former two by the binomial function and the latter by combinatorics or the multinomial. With two dice, diagrams can solve this type of probability question, e.g. *'a 4 with any other number'* for 1d6 with 1d4 = $^8/_{24}$ (Fig. 18.5) and counting from lists and diagrams, as was done in the example above with 3d6 (Fig. 18.4).

For direct calculation, the individual die probabilities require amendment to take into account the unnamed nature of some of the dice. With named identified outcomes there is but a single outcome (hence the $^1/_6$ probability for d6). For the non-specific cases, there are up to 6 possibilities with a d6, but this must be reduced to avoid infringing any conditions or *constraints*, applied to the event, ranging from all possible, downwards.

Some of the problems listed above, where we require the probability of the *content*, with one constraint only for the unnamed, can be solved using the ***binomial formula***.

The binomial method for dice

Dice events of this form pose an exact level of occurrence, or a range, for a particular die face value on one or more dice, e.g. the probability of *'getting two 6s on a roll of 5d6'*, *'getting zero 9s when I roll 2d16'*, or *'When I roll 5d12 what is the probability of getting at least three 11s?'* The method can handle single dice but probabilities are simpler here and the single dice methods (Section 18.2) are convenient. As seen, a range or inequality can be specified for the success circumstances, but this can also be done for the number of face values within the success probability, as in *'getting 4 or more on each die when 3d6 are rolled'*, etc. Any unallocated dice are allowed to take *any value except for the one specified*. This means that distinct or repeated outcomes will all be counted in the binomial calculation, Thus, for the event *'getting two 6s'* on 5d6, outcomes such as '6,6,2,5,1', '6,6,4,4,5' and '6,6,3,3,3' all qualify.

The binomial method is explained in Chapter 9 (Book One Section 9.2) and for coins (Section 16.4.2). In Excel, this is implemented with the

BINOM.DIST( ) function, amended for dice:

$$P_{dice\ content} = BINOM.DIST(k,\ n,1/s,\ 0\ or\ 1)$$

Where, k = no. successes, n = no. of dice (throws), s = dice size **(18.8)**

Thus, *'getting exactly three 7s with 5d8'*,

P = BINOM.DIST(3,5,1/8,0) = 0.0149536 (1%)

The event *'getting zero 9s with 2d16'* gives rise to some unusual detail:

P(*'x,y'* where x,y = 1-16, except 9')

= BINOM.DIST(0, 2,1/16, 0) = 0.878906 (88%)

We have a very good chance of not getting a '9' on 2d16!

When the binomial function is applied in this manner, the probability is the same for all other similar events. For example, in the former case, the probability is identical to that for 'three 1s', 'three 2s', ... 'three 7s' and three 8s'. So, to get a more general event, such as *'any three of a kind with 5d8'* the probability is multiplied by the number of outcomes on a single d8:

P(*'any 3 of a kind, 5d8'*) (can include pairs)

= 8 * BINOM.DIST(3,5,1/8,0) = 0.119629 (12%)

Ranges and Inequalities for dice with binomial

Game players are more likely to want to know ranges of specific values, such as an event of *'at least two 6s on 5d6'*. To get this probability the long way, probabilities for 2, 3, 4 and 5 '6s' must be added, giving 0.196. Or, we can use the complement, where,

P(*'at least two 6s'*) = 1 − P(*'less than two 6s'*),

but we are still required to do some addition for 0 and 1 '6s'. The binomial function is far easier, by providing an *inequality* for the number of successes as described for coins, so,

P = 1- BINOM.DIST(2-1,5,1/6,1) = 0.196245 (20%)

Similarly, *'getting at most four 2s with 6d4'* ,

$$P = \text{BINOM.DIST}(4,6,1/4,1) = 0.995361 \ (99.5\%)$$

This one is almost certain - it is saying that you are very unlikely to get more than four '2s' on throwing 6d4.

'getting at least three 11s with 5d12'

$$P = 1 - \text{BINOM.DIST}(3-1,5,1/12,1) = 0.00508777 \ (1\%)$$

We can also use the addition rule (formula (5.2b)) for non-mutually exclusive events of this type:

'at least one 6 on 3d6',

$$P = ('6 \ on \ d1 \ OR \ 6 \ on \ d2 \ OR \ 6 \ on \ d3') - 3 *('6 \ on \ 2 \ dice') + ('6 \ on \ 3 \ dice')$$

$$= \frac{1}{6} + \frac{1}{6} + \frac{1}{6} - 3 * \frac{1}{36} + \frac{1}{216} = \frac{91}{216}$$

$$= 1 - \text{BINOM.DIST}(1-1,3,1/6,1) = 0.421296 \ (\frac{91}{216}, \ 42\%)$$

The event in its full form is *'one 6 on 3d6 or two 6s on 2d6 or three 6s on 3d6'*. There are three 2-way joint events and one 3-way (see Book One Section 5.2.2). The function is of course easier to implement but it is informative example of non-mutually exclusive events.

A range can be included for the success probability, thereby increasing the possibilities, as with *'a 4 or more on each die for 6d6'*. This requires changing the individual die success probability value (p). For a d6, there are three values which qualify for success ('4, 5 and 6'), so $p = \frac{3}{6}$,

$$P = \text{BINOM.DIST}(6,6,3/6,0) = 0.125000 \ (12.5\%)$$

Note that the 'fail' probability would also be $\frac{3}{6}$ and a fail would generate face values other than those specified – in this case '1, 2 or 3' (see explanation Book One Section 9.2).

Similarly, *'getting exactly three instances of 7- 9 inclusive with 5d14'*

P = BINOM.DIST(3,5,3/14,0) = 0.0607448 (6%)

The number of successes is required as an exact number but the probability is entered to include outcomes '7, 8 and 9'.

Both inequalities can be included and the successes and the probability cover a range:

'getting one or more 7s or more with 6d10'

P = 1- BINOM.DIST(1-1,6,4/10,1) = 0.953344 (95 %)

A 'excellent' chance, but it does include '7s, 8s, 9s and 10s'.

Several of these example dice throws are found in game systems, such as war games and role-play or adventure, where a target value is designated and players attempt to get one or more dice to qualify. A more illustrative example would be in a war game - a hit on enemy units is gained when for each attacking unit a '5' or a '6' is required - if 7 units attack, what is the probability of 4 or more of the attackers scoring a hit? This means *'at least 4 hits'* (successes), each with a probability of $^2/_6$ (for the '5' or '6'),

P = 1 - BINOM.DIST(4-1,7,2/6,1) = 0.173297 (17%).

This particular use ('at least' with extended probability) of the binomial function is applied in calculating some probabilities in the game *Risk* ® (Section 20.2).

Dice events with named and unnamed outcomes

Where there are more stringent constraints on the unnamed or there are mixed outcomes of the sort *'2,5,x,y'* where x,y≠2,5 and *'4,4,x,y'*, where x≠y≠4, the binomial would be unsuitable. Use could not be implemented in the first example as one of the requirements is violated – the probability of failure (i.e. not getting a '2' (or a '5')) is not equal to 1- p. Applying the function to the second would result in over counting as occurrences such as

'4,4,5,5', '4,4,2,2', etc. would be included. That is, with the binomial method the unnamed is allowed to take repeated as well as distinct values. For games, there can be other particular constraints, such as excluding certain face values in some probabilities for *Farkle*, or one face can cancel another. At least three other methods are possible for these events:

- Direct calculation by individual probability for each die
- Calculate a particular case first then generalise
- Multinomial use

Generally, each calculation route must consider distinct and repeated face values where appropriate, in both named and unnamed parts of the configuration.

Direct calculation individual probability

This deals with individual dice one at a time, to give the 'in order' probability first. With the named part, probabilities are all for the single die outcome,

$P(named+unnamed)_{in order\ direct}$:

$= in\ order\ probability\ named\ (direct\ calc.)$

$* ind.\ prob.\ for\ unnamed\ within\ constraints$ **(18.9a)**

e.g. in '2,5, x,y' the '2,5' part has probability $= \frac{1}{6}^2$

For the unnamed, any constraints need to be declared – such as 'x,y' cannot equal '2 or 5' and they cannot equal one another, resulting in the net probability,

$P('2,5,\ x,y'\ in\ order,\ x{\neq}y{\neq}2,5') = \frac{1}{6}^2 * \frac{4}{6} * \frac{3}{6} = \frac{12}{6}^4 = 0.00925926\ (1\%)$

There are 4 ways to choose the first unnamed, leaving 3 for the second. Note that 'in order' applies to the named part only and 'any order' to the unnamed, as this latter part can, depending on constraints, have distinct or repeated values.

This method is simple but as seen, it introduces complications when we move to the 'any order' state for the whole sequence. The individual probability calculation for the unnamed generates all its arrangements. In some cases, the unnamed value count is analogous to a lesser number of smaller sized dice. To demonstrate this, take the two unnamed above that, because of the constraints can take 3 and 4 values - this is equivalent to two distinct faces on 2d4, giving $^4P_2 = 12$ arrangements ($= 3*4$ as above). Formulae below (Section 18.4), are useful for enumeration of, or for a check on, this counting problem with distinct and repeated unknown sequences.

We have the 'any order ' for the unnamed and to integrate this with the named part, a slightly convoluted process is required. Each of the unnamed sequences can be formed up with the named outcomes, by 'slotting them in' at the number of locations that they can take, using combination formula (7.5a), (^nC_r; Book One):

P(named+unnamed)$_{anyorder\ direct}$**:**

$= $ *in order probability (direct calc.)*

* *ways for named to slot in*

* *arrangements of named*   **(18.9b)**

We get the orders by use of permutation formulae for all distinct and repeated for the named part. Take the '2,5,x,y' example, where unnamed and named both exceed 1, the named outcomes can be meshed in with the unnamed, i.e. slotted in as '4 taken 2 at a time' ($^4C_2 = 6$), so that they can occupy locations 1,2/ 1,3 /1,4, /2,3,/ 3,4 and 2,4. So, '2,x,5,y' would be one of the outcomes. At the last step, we see that '2,5' forms 2! arrangements using the n! permutation formula. This gives $2*12 = 24$ arrangements in total, half with '.2..5.' within the sequence and the other 12 with '.5..2.':

$P('2,5,x,y',any\ order;x{\neq}y{\neq}2{\neq}5) =$

Inv. prob.
in order   slot in   named arr.

$= {}^{12}/_{6}{}^{4} \qquad * \ {}^{4}C_2 \qquad * \ 2!$

$= 12 * 6 * 2 \ /6^4 = {}^{144}/_{1296} = 0.111111 \ (11\%)$

Direct calculation - repeated face values in named and unnamed

The former case, i.e. **repeats in the named**, is treated in the same way, adjusting for the values available for unnamed:

$P('5,5,x,y',any\ order;x{\neq}y{\neq}5) =$

In order probability   slot in   named

$= {}^{1}/_{6}{}^{2} * {}^{5}/_{6} * {}^{4}/_{6} \qquad * \ {}^{4}C_2 \quad * \ 2! \ /2!$

$= 20 * 6 * 1 \ /6^4 = {}^{120}/_{1296} = \ 0.0925926 \ (9\%)$

Similarly for more unnamed, $'3,3,x,y,z'$ where $x{\neq}y{\neq}z{\neq}3$,

$P = {}^{1}/_{6}{}^{2} * {}^{5}/_{6} * {}^{4}/_{6} * {}^{3}/_{6} = {}^{60}/_{7776}$ in order and

$P = 60/7776 * {}^{5}C_3 * 3!/3! = {}^{600}/_{7776} = 0.0771605 \ (8\%)$, any order

This is the probability of a particular pair and to get 'any pair' for *Poker Dice* and similar games, we multiply by the number of possible pairs with 6 digits (1,1/ 2,2/ 3,3 …6,6) so:

$P('any\ pair\ on\ 5d6') = {}^{600}/_{7776} * 6 \ = {}^{3600}/_{7776} = 0.462963 \ (46\%)$

For cases where the **unnamed** can have repeats (pairs, trebles, etc.) or where named = unnamed, calculation is best left to the other methods rather than direct calculation as with other than simple cases it gets complicated with confusion of configurations and their arrangements (Book One Section 13.3; see below and Problems & Exercises).

Specific case

This method overcomes the above complexity. The original problem is converted to a specific case within the required constraints, so,

$P(named+unnamed)_{anyorder\ sp.\ case}:$

= *specific case for all face values*

    * *general case combinations for unnamed*

    * *factorial arrangements for all*   **(18.9c)**

Thus, for '$2,5,x,y$' with $x{\neq}y{\neq}2,5$' means that all elements are *distinct* and a specific case or instance would be the sequence '$2,5,1,4$', notated as,

    '$2,5,x,y$' distinct $\rightarrow$ '$2,5,1,4$'

We apply (18.9c) to this and,

$$P = \frac{1}{6}^4 * {}^4C_2 * 4! = 144/6^4 = 0.111111$$

The 'in order' probability for the specific sequence is simply as formula (18.5a). The generalisation of this for '<u>any x,y</u>', is a selection of 2 from 4 numbers (1,3,4,6), by the combination formula, which does not generate arrangements. To get any order, we simply multiply by the appropriate permutation formula in this case, n! where n = 4.

Formula (18.9c) accommodates **repeats** in named, as,

    '$2,2,x,y$' $\rightarrow$ '$2,2,1,4$'

$$P = \frac{1}{6}^4 * {}^5C_2 * 4!/2! = 10 * 12/6^4 = {}^{120}/_{1296} = 0.0925926\ (9\%)$$

This is similar to the above except that there are now 5 faces available for selecting the 2 unnamed distinct values.

For **repeats** in **unnamed**, '$x{=}y$ but $\neq 2,5$'

    '$2,5,x,y$' repeated $\rightarrow$ '$2,5,4,4$'

$$P = \frac{1}{6}^4 * {}^4C_1 * 4!/2! = 48/6^4 = 0.0370370\ (4\%)$$

(The unnamed are $\equiv$ any pair on 2d4 = 4)

If we wish the event to include both distinct and repeated, then these probabilities are added (P = $^{192}/_{1296}$ = 0.148148 (15%)).

Multinomial

Using multinomial formulae follows as similar path to the above, by specifying a particular case, calculating the combinations for the unnamed, then generalising if appropriate as described in Book One Section 11.2:

P(named+unnamed)_{anyorder multinomial}:

P = *specific case for all * multinomial general case for unnamed*

* factorial sequences for all   **(18.9d)**

Multinomial use applies to the <u>unnamed</u> only, thus for the *'2,5,x,y'* example there are 2 instances of the unnamed and a choice of two values from four,

'2,5,x,y' distinct → *'2,5,1,4'*,

1 3 4 6

1 0 1 0   = MULTINOMIAL(2,2) = 4!/(2!2!) = 6 configurations

'2,5,x,y' → *'2,5,1,4'*

1 3 4 6

1 0 1 0   = MULTINOMIAL(1,0,1,0) = 2!/(1!...) = 2 arrangements

This covers the unnamed only and the two named values need to be slotted in with their arrangements,

$^{4}C_2$ * 2! = 12 and the net count = 6 * 2 * 12 = 144

Therefore, 'x,y' can occur as 6 distinct configurations, each with 24 sequences for the whole event. (Alternatively use the permutation formula as 4! to get the number of arrangements). A similar multinomial treatment can be given for x = y. The multinomial calculator, highlighted below, can be used for these problems but data must be entered adjusted for the number of elements in the unnamed.

Using these methods covers any 'named + unnamed' problem for different numbers, each with distinct and repeated instances. The examples in

Problems & Exercises illustrate some of the variations.

Named and unnamed application

The most common application is for calculation of probability for scoring patterns as *content* in dice games such as *Poker dice* and *Yahtzee®*, as illustrated by one example above and below (Section 20.2). The specific case is determined then this is extended to cover all forms as shown above.

There are also circumstances where there are no named dice face values as such. This is where a general pattern is specified such as '*4 oak and 2 other distinct values*', etc. The specific case or multinomial method can be used, but an additional multiplier is required in the former for the number of repeated forms possible with the size of the dice in the event. The multinomial gives the answer more directly and the calculator below (Spreadsheet table 18.2) can do the calculation after simple input of a specific case (Section 18.4), e.g.,

'*a pair and a distinct face value*' = '*x,x,y on 3d6* ' where $x \neq y$

   Specific case → '*3,3,6*',

     $P = 1/6^3 * {}^5C_1 * 3!/2! = 15/6^3$

This covers all possible values for the single distinct face, but only deals with one of the possible pairs, Thus for all pairs,

   $P = 15 * 6 /6^3 = 90 /6^3 = 0.4166667$ (42%)

Another example using the multinomial method,

'*a pair and a treble on 5d6*' '*x,x,y,y,y*' → '*5,5,1,1,1*',

1 2 3 4 5 6
3 0 0 0 2 0 = MULTINOMIAL(4,1,1) = 6!/(4!1!1!) = 30 configurations

→ '*5,5,1,1,1*',
1 2 3 4 5 6
3 0 0 0 2 0 = MULTINOMIAL(3,0,0,0,2,0) = 5!/(3!2!)

$$= 10 \text{ arrangements, net} = {}^{300}/_{7776} = 0.0385802 \text{ (4\%)}$$

These methods were used to prepare a table (Table18.2b) of various patterns for numbers of d6.

With dice and unnamed outcomes, the 'in order' probabilities are useful for game situations when two or more players are rolling dice and where the order can be allocated player-wise. An example of this enumeration was illuminated in Section 15.2.1 for two players. Now take three players (A,B,C as d1,d2,d3) who roll a d6 one after the other. If the winner is the one who gets a '5' and the others do not, we can apply the unnamed procedures to calculate any player's chance,

A win - all sequences '5,x,y' where x,y≠5

$$P = {}^{1}/_{6} * {}^{5}/_{6} * {}^{5}/_{6} = {}^{25}/_{216}$$

B win roll must be 'x,5,y' where x,y,≠ 5,

$$P = {}^{5}/_{6} * {}^{1}/_{6} * {}^{5}/_{6} = {}^{25}/_{216}$$

C win roll must be 'x,y,5' where x,y,≠ 5,

$$P = {}^{5}/_{6} * {}^{5}/_{6} * {}^{1}/_{6} = {}^{25}/_{216}$$

Thus, each player has an equal chance. There are 75 sequences where a 'win' event can occur and the other 141 where there is 'no winner' and the cycle repeats to infinity. A geometric series once again and following the steps to sum the series (Section 15.2.1):

1st turn A win = ${}^{25}/_{216}$

2nd turn A,B,C fail then A win = ${}^{141}/_{216} * {}^{25}/_{216}$

3rd turn (A,B,C fail) (A,B,C fail) then (A win) = ${}^{141}/_{216} * {}^{141}/_{216} * {}^{25}/_{216}$

OR ${}^{141}/_{216} * {}^{141}/_{216} * {}^{141}/_{216} * {}^{25}/_{216}$

OR ... ∞,

First term = ${}^{25}/_{216}$, common ratio = ${}^{141}/_{216}$

$$P('A \text{ win}') = {}^{25}/_{216} / (1 - {}^{141}/_{216}) = {}^{25}/_{216} * {}^{216}/_{75} = {}^{25}/_{75} = {}^{1}/_{3} = 0.333333 \text{ (33\%)}$$

This probability applies to B and C also. Just like previous examples in Section 15.2.3, there is no advantage by going first. As 'no win' turns can occur, bets could be placed and allowed to accumulate until a win happens.

Win probability can be calculated for any number of players, e.g. for two players, win for A = '5,x', $P = \frac{1}{6} * \frac{5}{6} = \frac{5}{36}$ on single turn for both, but for a series $= \frac{5}{36} / (1 - \frac{26}{36}) = \frac{5}{36} * \frac{36}{10} = \frac{1}{2}$. Note that these results differ from those obtained in Section 15.2.3, which used formulae, but included ties.

Multiple dice - runs of face values

Just like coins, we can have a *run* of a particular face value with dice, e.g. 'a run of four 5s with 6d6'. This is a consecutive sequence, such as *'(5555)'*, which can occur on its own, an 'isolated run sequence' or along with fails or with mixtures of fails and successes (as described in Section 15.2.3 & Book One Section 5.3.3) and Section 16.4.2 for coins. With dice the 'success' is identified as a single face value, but the 'fails' can be represented by several possibilities, e.g. if the success is '7' on d10 then the fail outcome can be '1-6 or 8-10', etc. The run can be depicted in the form above along with, in the example, two other face values, in the style *'(5555),x,y'* ('x,y' in this case denote where the base action fails). The probability for an isolated dice run sequence is calculated by p^n where p is the probability for the designated face outcome. This in turn depends on the dice size (s) and the number (n) of dice or throws, hence, $P_{isolated\ dice\ run} = 1/s^n$. So, for *'a run of five 7s on 5d8'*:

$$P('(77777)\ on\ 5d8') = \frac{1}{8}^5 = 0.0000305176$$

For the *simple run* sequence, where the run contains all the successes separate from a number of fails, can be determined most conveniently by the expressions of Chapter 5, although the named + unnamed methods above are also possible. Using formula (5.3) (Book One) for a simple run:

$$P = (f + 1) * p^r * q^f$$

Thus, for '66x' for 3d6, f (no. fails) = 1, r (length) = 2, p(success) = $\frac{1}{6}$, q(fail) = $\frac{5}{6}$ then,

$$P = 2 * \frac{1}{6}^2 * (\frac{5}{6})^1 = \frac{10}{216} = 0.0462963 \ (5\%)$$

By the unnamed methods, the run must be uninterrupted and is treated as one unit, i.e. '(66),x' ≡ '6,x', constraint x≠6, having 1 named and 1 unnamed, then calculated using direct calculation as:

'(66),x' with 3d6, $P = (\frac{1}{6})^2 * \frac{5}{6} *^2C_1*1! = \frac{10}{216}$

This tells us that there are 10 ways to get a run of '66' and we can refer to Figure 18.4 and count these instances to confirm (visually in this case, with '66' marked in bold for counting). (Note that this run event is not the same as P('*content* = '6,6,x') = $\frac{15}{216}$)

With more fails such as '(5555),x,y' on 6d6,

$$f = 2, r = 4, p = \frac{1}{6}, q = \frac{5}{6}$$
$$P = 3 * (\frac{1}{6})^4 * (\frac{5}{6})^2 = \frac{75}{46656} = 0.00160751 \ (0.2\%)$$

Similarly, '191919,x,y' on 5d24', (Table 18.1):

$$f = 2, r = 3, p = \frac{1}{24}, q = \frac{23}{24}$$
$$P = 3 * \frac{1}{24}^3 * (\frac{23}{24})^2 = 0.000199306 \ (0.02\%)$$

Another layer of complexity arises if we also allow other occurrence of the run value as a 'complex run' where the sequences can have mixtures of fails and success and inclusion of longer and multiple runs. As explained, these are beyond the scope of this text, other than by counting from lists or diagrams or by computer simulation (Book One Section 14.3). Some examples by this latter method are,

10d6 - run of 5 (particular face value), an estimate,

P('*complex run of 5 or more*') = 0.000694

However, in this case only 1 run of length 5 -10 is possible. As seen by the latter result and others above, some run occurrences are rare events.

Probabilities can be enhanced by inclusion of 'any run' rather than a particular one, e.g. P(*'any run of 5, 10d6'*) = P(*'particular run of 5'*) * 6 = 0.4%.

Nevertheless, some dice runs can be important. During a role-play game d20 is used for combat and a typical encounter involves 67 rolls of 1d20 – what is probability of a simple and a complex run of 'three 20s' (an especially effective hit)?

P(simple run) = $(f + 1) * p^r * q^f$ = 65 * $^1/_{20}$^{3} * $(^{19}/_{20})^{64}$ = 0.0003 (<0.1%)

P(complex run at least 1 of at least length 3 by simulation) = 0.0072 (1%)

Not very likely for a simple run, but with other runs of 20s appearing in the complex format, the probability is higher by a factor of 20 or more. Three consecutive appearances of '20' by one opponent or group can have dramatic effects on the result of combat encounters in such games. Any run of '3 of a kind' is also more probable and within real games a run of length 3 crops ups about 1 in 100 rolls (see more examples in games below).

Expected value and variance for multiple dice

Single dice expected value and variance were dealt with in Section 18.2.1. This time a number of different types of occurrence are considered. For dice face value outcomes that can be modelled by the **binomial distribution**, we can use the parameters (n*p, n*p*(1-p), respectively) of that distribution (Book One Section 9.2). Thus, for a case where the random variable is 'the number of 6s on throwing of 5d6', n = the number of throws (5) and 'p' is the probability of the individual throw ($^1/_6$), then:

E(X) '6s on 5d6' = 5 * $^1/_6$ = 0.833333, i.e. the expected number of '6s' ≈ 1

V(X) = 5 * $^1/_6$ * (1- $^1/_6$) = $^5/_6$ * $^5/_6$ = $^{25}/_{36}$ = 0.694444, SD = 0.833333 ≈ 1

Thus, the average and standard deviation of 5d6 have the same value. Consequently, when throwing 5d6 repeatedly, the average would be one '6' ranging from 0 – 2, and this applies to any other face value. The mode, or rather one of them, is also 1 and has the highest probability (40%; other mode is 0).

Likewise for other numbers and sizes of dice,

$E(X)$ '1s on 7d10' $= 7 * {}^1/_{10} = 0.7 \approx 1$ (mode 0)

$E(X)$ '5s on 48d6' $= 48 * {}^1/_6 = 8$ (mode 8)

These $E(X)$ counts are the <u>average no. of outcomes</u> with a particular face value.

Average number of throws and success by nth trial with multiple dice

Distinct from the above, there is another count of interest: the <u>average no. of throws</u> to get a particular multiple outcome of face values. For this, we require the probability of the event followed by calculation of the *geometric distribution* expected value (1/p; as done for single dice above).

For example, to get the average number of rolls for 'box cars' in *Craps* on 2d6 - the random variable is the no. of throws 1,2,3, ... and,

$P('2d6, box cars') = {}^1/_{36}$ $E(X) = 1/({}^1/_{36}) = 36$

Contrast this with a '7' to win, $P('2d6, '7') = {}^6/_{36}$ $E(X) = 1/({}^1/_6) = 6$

Larger dice - *'any three of a kind with 3d8'*

$P('3d8, three of a kind') = {}^8/_8{}^3$ $E(X) = 1/({}^1/_{64}) = 64$

With *inequalities*, such as *'at least four 1s on 10d12'*

$P = 1 - BINOMIAL.DIST(4-1,10,1/12,1) = 0.00671625$ (1%)

$E(X) = 1/0.00671625 = 148.892593 \approx 149$

Therefore, a long time for this one, with the average at almost 149 throws.

Also, used for sums of multiple dice:

$P('sum of 4 on 2d6') = {}^3/_{36}$ $E(X) = 1/({}^1/_{12}) = 12$

A dozen throws on average.

We can also calculate the expected value for an event over a specific number of repeats of an experiment. If 5d6 are used and the event is *'three 6s'*, P = 0.0321502 (from binomial). If we throw 5d6 one hundred times then the number of appearances, on average, of the content of exactly 'three 6s' would be,

E(X) 'three 6s, 5d6, 100 throws' = 100 * 0.0321502

= 3.215 ≈ 3 times ('3 or more' ≈ 4 times)

The expectation for multiple outcomes can be obtained via the E(X) of the multinomial distribution as per the example in Ch 11 on the dice game *Farkle,* (Book One Section 11.2). The expectations for '1s' and '5s' (both scoring faces) on a roll of 6d6 are both 1, but this is not necessarily for both together (probabilities for occurrence of '1' and '5' once, twice and three times are = 16%, 3% and 0.04 %, respectively, Section 20.2.1).

In the game of *Crown and Anchor* on 3d6, expectation for the number of 'crowns' is,

P(*'crown'*) = $^1/_6$ E(X) 'crown' = 3 * $^1/_6$ = 0.5

We can confirm this figure by using formula (8.1a) Book One Section 8.2.2,

E(X) 'crown' = 1 * $^{75}/_{216}$ + 2 * $^{15}/_{216}$ + 3 * $^1/_{216}$ = 0.5 (mode 0 at 58%)

The same calculation applies to 'anchor ' so on average these values appear once every two throws (probability for both together is 11% (Section 20.2.1)).

The 'how long before success' expected values and probabilities for success(es) at any stage are accessed by geometric and negative binomial functions. To calculate the probability for *'three 3s on 3d6 by 3rd throw'*, we view the throw of 3 dice as a compound event that is repeated, then,

P(*'success on 3rd multiple throw'*) = NEGBINOM.DIST(2,1,1/216,0)

$$= 0.00458686\ (0.5\%)\ = (215/216)^2 * 1/216$$

This is not the same as *'getting three 3s by the 3rd roll'*, where a single die is thrown three times,

$$P = \text{NEGBINOM.DIST}(0,3,1/6,0) = 0.00462963\ (0.5\%) = (1/6^3)$$

For <u>mixed dice</u>, the multiple throw is viewed as the overall event and it is repeated, so,

'Probability of waiting until the 10th throw until get three 1s on roll of 1d10, 1d12 and 1d14':

$$P = \text{NEGBINOM.DIST}(9,1,1/(10*12*14),0) = 0.000592057\ (0.1\%)$$

18.4 Fuller characterization of dice outcomes

The total outcomes of multiple dice face values can be partitioned into the numbers of permutations (sequences) and combinations (configurations) with distinct values and repeats. The sequences can be further partitioned into those containing single and multiple repeated elements (doubles, trebles, etc.). The formulae above are used and some are summarised below (with the Excel version following):

Total no. sequences (with and without repeats) = s^n (= s^n)

No. distinct sequences = ^sP_n (= PERMUT(s,n))

No. of sequences with repeats = $s^n - {}^sP_n$

No. distinct <u>content</u> combinations (without repeats)
   = ^sC_n (= COMBIN(s,n))

No. content combinations with and without repeats
   = $^{s+n-1}C_n$ (= COMBIN(s+n-1,n))

No. of content combinations with repeats = $^{s+n-1}C_n - {}^sC_n$

   Where, s = dice size and n = no. of dice

   Content combinations ≡ configurations

s<= n in COMBIN( ) and PERMUT( )

(i.e. cannot get 7 distinct face outcomes with six-sided die, etc.)

The expressions were used to prepare Table 18.3a for 2 to 6d6. These counts can aid calculation of probabilities but care is required with combinations with repeats as these do not all occur in the same frequency. The number of distinct configurations falls to 1 when s = n = 6 (this would be so for any similar circumstance, e.g. 8d8).

Using these formulae and others (the multinomial calculator in particular), a companion table (Table 18.3b) was constructed displaying counts for various forms of sequence, mostly for repeated elements. The exception is for all outcomes to be distinct, where PERMUT(s,n) is used to give 'singles'.

Table 18.3a Partitions of dice sequences and configurations

	No. of	2d6	3d6	4d6	5d6	6d6
s^n	total seq.	36	216	1296	7776	46656
^sP_n	distinct seq.	30	120	360	720	720
$s^n - {}^sP_n$	repeat seq.	6	96	936	7056	45936
^sC_n	distinct conf.	15	20	15	6	1
$^{s+n-1}C_n$	total conf.	21	56	126	252	462
$^{s+n-1}C_n - {}^sC_n$	repeat conf.	6	36	111	246	461

The latter table (18.3b) allows quick access to solutions and checks for problems in outcomes in dice games (Section 20.2). To get probabilities, the count from the table is divided by the total number of sequences (top of the table) as per the classical formula (3.1) (Book One).

Thus, *'what is the probability of getting a double treble in the dice game Farkle (6d6)?'*

$$P(\text{'2 trebles on 6d6'}) = \frac{300}{46656} = 0.00643004 \ (1\%)$$

The tables cover up to 6 dice but they can be extended for more and modified for different sizes.

Six-sided dice are covered, but it is useful to calculate for other sizes in themselves and as an aid in problems with unnamed dice, which can have configurations based on smaller dice. For example, where three unnamed cannot have one face value of a d8, they would be based on the sequences in 3d7.

Table 18.3b Partitioning of sequences in dice distributions

	No. of d6					Example sequence (6d6)
	2	3	4	5	6	
Total	36	216	1296	7776	46656	-
Singles	30	120	360	720	720	3,2,4,1,6,5
Pair	6	90	720	3600	10800	5,5,3,1,6,2
Treble		6	120	1200	7200	2,2,2,5,3,1
2 Pairs			90	1800	16200	4,4,1,1,5,2
Quad			6	150	1800	6,6,6,6,3,4
Treble + Pair				300	7200	3,3,3,5,5,2
Quintet				6	180	1,1,1,1,1,4
3 Pairs					1800	5,5,2,2,6,6
Quad + Pair					450	4,4,4,4,5,5
2 Trebles					300	6,6,6,1,1,1
Sextet					6	3,3,3,3,3,3
Total	36	216	1296	7776	46656	

Any particular calculation can be performed by several methods as described above, but some more general formulae are possible, with the multinomial method arguably the most versatile. Where one face value is repeated, one general formula can be used,

No. of sequences with one repeated face value:

$$= {}^sP_{n-rpt+1} * {}^nC_{(n-rpt + (rpt<2))}$$

$$= PERMUT(s, n - rpt +1) * COMBIN(n, n - rpt + (rpt<2))$$

Where, n, s = no. and size of dice, (s=< (n-rpt+1))
    rpt = no. of repetitions, 1 = none (all distinct)
        2 = pair, 3 = treble, etc.   **(18.10)**

For example, *'how many sequences with a treble can be thrown with 4d6?'*

No. seq. with treble, 4d6:

$n = 4$, $s = 6$, $rpt = 3$, $(rpt<2) = FALSE = 0$

$= PERMUT(6,4 - 3 +1)) * COMBIN(4,4-3 + 0)$

$= {}^6P_2 * {}^4C_1 = 30 * 4 = 120$ & $P = 120/6^4$

More dice can be used as with using the ${}^{s+n-1}C_n$ formula from Table 18.3a,

No. configurations for 7d6 $= {}^{6+7-1}C_7 = {}^{12}C_7 = 792$

If 'rpt' is set at '1', (18.10) reverts to the formula above for 'all distinct sequences' (${}^sP_n$). Some instances of repetition are not possible, e.g. with 8d6 it is not possible to get a single pair, so the calculation will give an error. This formula and others below can be used to explore probability for dice and some other devices) for the purpose of existing and new games. For example, the small output below shows the most probable number of repetitions for 6d6 to 10d6. Overall repetition probability falls with the number of dice (the possible levels of repetition depend on the numbers of dice). We can see that with a game based on 'getting two or more repeats' with different numbers of dice, probability would be counter-intuitive in a way because a player may conceive that 'more dice' means more chance of getting a repeat:

Spreadsheet Table 18.2

	6d6	7d6	8d6	9d6	10d6
Most probable	2 (23%)	3 (9%)	4 (3%)	5,4 (0.9%)	5 (0.3%)

For multiple repetitions such as *'a pair and a treble'*, it is possible to calculate by a single long expression for all forms with factorials, but this gets unwieldy as the number of repeats get larger and it is more convenient to use the multinomial method. This latter route is illustrated above and displayed in Spreadsheet Table 18.3.

Multinomial calculator

The sheet is set up for up to 10 dice of size up to 10. Three data items are entered, beginning with 'n' and 's' (number and size of dice), which are named as such using R-click Define Name. A specific case of the sequence of dice outcomes of interest follows. This can be up to 10 elements in length and it must match 'n' within 's' in magnitude. It is named as 'sequence' (i.e. this is an array 10 cells long, named with Define Name: block select the 10 cells then R-click). Ensure that any cell(s) not required are cleared with the Delete key. The row above the sequence array is filled (except for the first cell, which is set = 1) according to 'n' using'

IF(n >=2,2,"-") → IF(n >=10,10,"-").

The sheet then creates two new output rows based on the dual application of multinomial expressions described previously (Book One Section 11.2 & Section 18.3.4). These are also named:

MN_1 (10 cells long) is filled with COUNTIF(sequence, cell above) and the row above (except for the first cell) is filled according to 's' with,

IF(s>=2,2,"-") → IF(s>=10,10,"-").

The "-" characters will not be counted by COUNTIF( ).

MN_2 (11 cells) is filled with,

COUNTIF(MN_1,cell above) - (10 -s) and

the row above (except for the first cell set = 0), with,

IF(n+1 >=2,1,"") → IF(n+1 >=11,10,"-")

At the foot of the output, the variable counts are computed using the MULTINOMIAL( ) function, which refers directly to the arrays (MN_2 has an adjustment for when s=2 (coins)). Counts for arrangements and configurations are named as 'ag' and 'cg', respectively. Additional features can be added such as headings ('blank', 'x of a kind' (1oak, 2oak ... 10oak) and error checking to ensure that input of sequence matches the specified number and size of dice.

Spreadsheet Table 18.3

MULTINOMIAL CALCULATOR											
DATA:											
Dice size (s) =		10	No. dice (n)=		10						
			Die number								
		1	2	3	4	5	6	7	8	9	10	-
Specific sequence =		9	9	5	5	5	7	7	7	7	10	-
Calculations:				Dice size							
		1	2	3	4	5	6	7	8	9	10	-
MN_1 =		0	0	0	0	3	0	4	0	2	1	-
		blank	1 oak	2 oak	3 oak	4 oak	5 oak	6 oak	7 oak	8 oak	9 oak	10 oak
		0	1	2	3	4	5	6	7	8	9	10
MN_2 =		6	1	1	1	1	0	0	0	0	0	0
Results:											

		Formula	
arrangemnts. (ag) =	12600	=MULTINOMIAL(MN_1)	
configuratns. (cg) =	5040	=MULTINOMIAL(MN_2) - (s=2)	
sequences(count) =	63504000	= ag * cg	
General probability:			
	P =	0.006350	= count/s^n
		0.6%	
		2/315	

It should be noted that the calculator gives probabilities for the <u>general case</u>. If you enter a fully named sequence, it will give the probability for any configuration of that form.

The example shown in the sheet contains 4 states of repetition (1-, 2-, 3- and 4oak ('of a kind')) with 10d10. The first multinomial calculation gives the number of arrangements for the original sequence and the second gives the number of different forms or configurations (i.e. <u>any</u> pair, treble and four of a kind). The product of these produces the number of sequences for these states and it is divided by the total number of sequences to give the final probability as 'any 10d6 content' with the above.

The calculator covers any number of repeats within the data range and it can be used for coins ($\equiv$ d2) and it can be extended to any practical size, such as s = 37 or 38 for roulette. We can do quite complex sounding probability for problems with more dice e.g. *'what is probability of getting 5 doubles with 10d6?'* For this, enter a sequence of 10d6 containing 5 pairs such as '2,2,4,4,3,3,5,5,1,1' into the **sequence** row. The spreadsheet calculates that this has 113400 arrangements and occurs in 6 configurations with P = $680400/6^{10} = 0.011253 \approx 1\%$.

For the name with unnamed problems, only the unnamed part goes into the calculator. In the case of '2,5,x,y' (x,y$\neq$2,5) there are two slots for unnamed and 4 face values are available, so the settings are dice size = 4 and dice no. = 2. Any sequence can go in within these limits, either as two distinct values or as two repeated, giving counts of 12 and 4 respectively. These require to be slotted in with the named ($12 * {}^{4}C_2 * 2! = 144$ and $4 * {}^{4}C_2 * 2! = 48$, agreeing with above in Section 18.3.4 (see Section 20.2.1 *Farkle* examples).

Chapter 19
Dice sums

19.1 Introduction

Addition of the face values of two or more dice (two or more random variables) produces another discrete random variable: that of the *dice sum*. Single dice outcomes exhibit a uniform distribution (Book One Section 8.2) but the <u>sum</u> of two or more independent uniform variables results in distributions of different nature and properties.

19.2 Dice sum distributions

The random variable for these distributions is based on this addition of the individual dice elements in each sequence. When plotted, a characteristic peaked form is produced ranging from a triangular distribution for two dice, to increasing bell shaped forms (see figures below). These shapes can be compared with a uniform distribution of the same range (Fig. 18.2). These displays all require that the data are frequencies determined empirically or by counting with simple examples, or with the probabilities obtained via the classical formula. Otherwise, the number of outcomes is calculated by one or more of the methods described in Section 19.3.

The two dice sum

The distribution of the sum of two independent uniform variables forms a *triangular distribution*, typified by that of two dice of the same size (Fig. 19.1). For this chart, the probabilities were calculated by counting the outcomes for each sum, via the data of Table 4.5 (Book One) and operating the classical equation (3.1), e.g. there are 36 outcomes in the sample space and *'a sum of 4'* occurs in 3 forms, hence P (*'sum 2d6= 4'*) = $^3/_{36}$ = $^1/_{12}$ = 0.0833333.

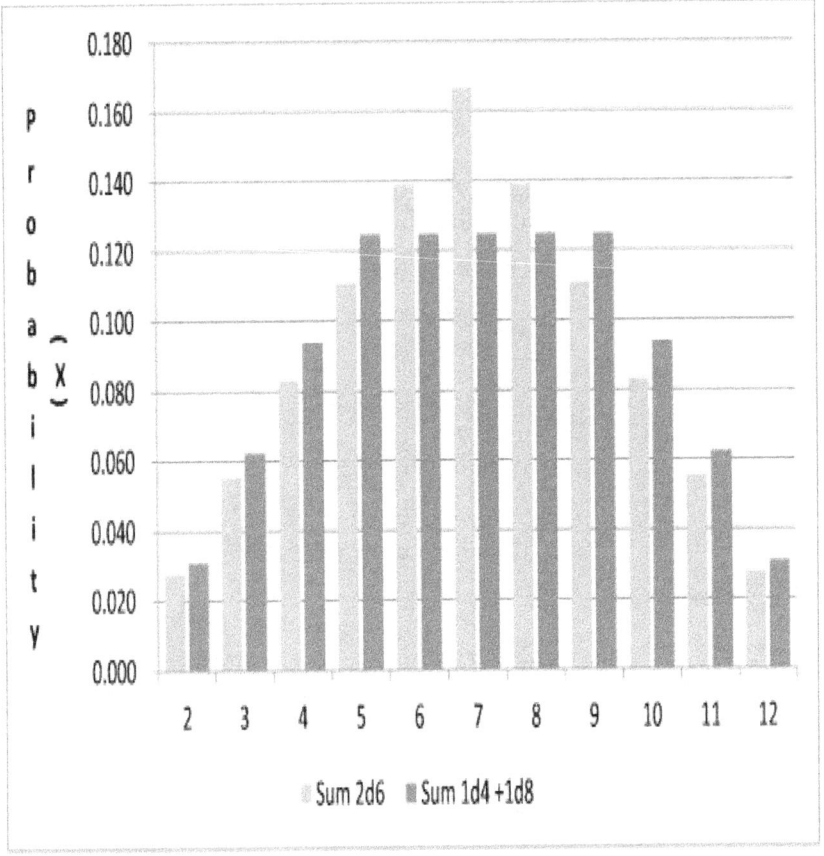

Figure 19.1 Probability distribution (2d6 and 1d4 + 1d8)

This display (2d6), is an instance of the ***discrete triangular distribution.*** It is recognized and defined and the two dice example gives a symmetrical form. The triangular shape applies to two variables with the same bounds, i.e. in this application, the same dice size. The sum of two dice of different sizes can show a variety of forms, characterised by a multimodal appearance, as also shown in the chart for 1d4 + 1d8. Thereafter, as the numbers of dice rise, sum distributions increasing approximate the normal distribution (Fig. 19.2).

Some mixed dice sums show similar distributional forms to those with only one size, e.g. the sum of 1d4, 1d6 and 1d8 has a close resemblance to 3d6 (Fig.19.3). Others such as d4 + d6 + d20 and 3d10, differ markedly (Fig.

14.3, Book One), as probabilities below 12 and above 21 are higher than those of the latter dice, and between this range the 3d10 ones become dominant. Such mixed dice distributions show that in many cases, sums cover the same range as those with one size, but have different probabilities, which could be useful in some games.

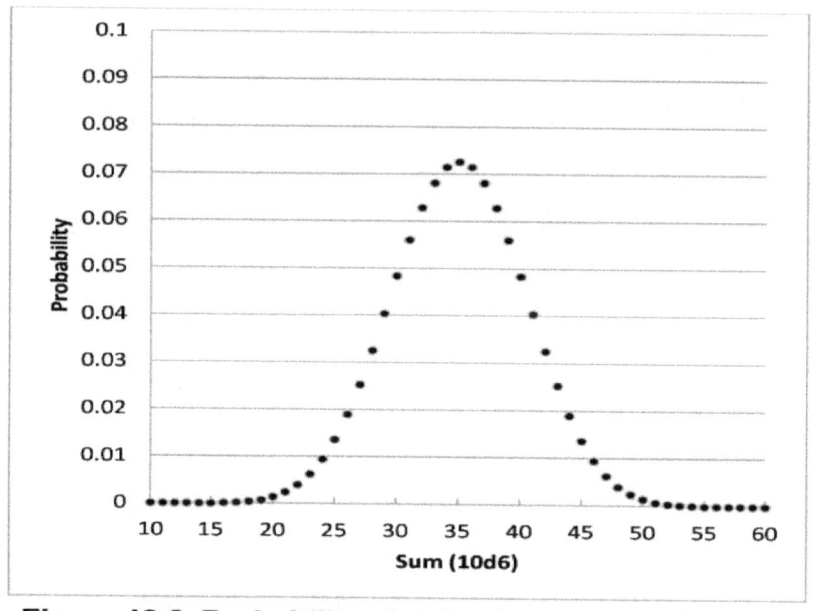

Figure 19.2 Probability distribution of sums for 10d6

19.2.1 Expected value, variance and probability function of dice sums

With other distributions, we were able to establish certain parameters that eased some probability calculations. With dice sums, we do not have a *general* distributional form and probability function that can be applied in a similar manner, although the triangular distribution parameters can be used for some with two dice.

The $E(X)$ and $V(X)$ parameters for multiple dice sums are obtained by simple addition of the corresponding values for single dice (Book One Section 8.2.2). Thus, for six-sided dice, the expected value for the sum is based on that value for 1d6, e.g. 3d6:

E(X) 'sum of 3d6' = 3 * E(X) '1d6 ' 3 * 3.5

$$= 3.5 + 3.5 + 3.5 = 10.5 \text{ (mode at } 10,11, 12.5\%)$$

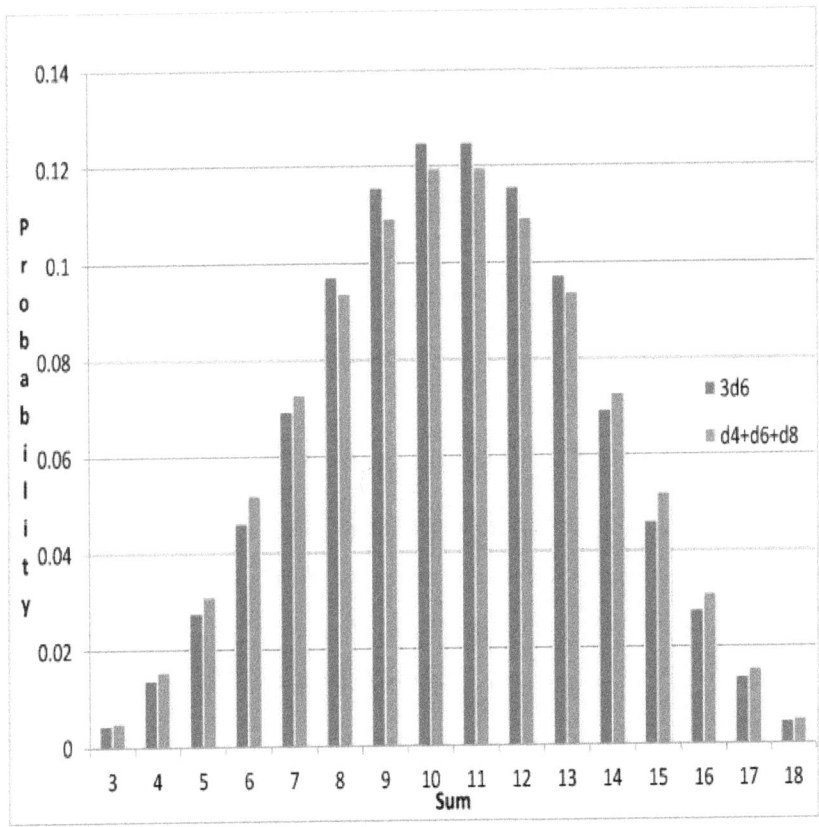

Figure 19.3 Probability distribution of sums for one size of dice (3d6) and for mixed sizes (d4+d6+d8)

This can be confirmed by summing the product of each outcome and its probability (8.1b):

$$3 * {}^1/_{216} + 4 * {}^3/_{216} + ... + 18 * {}^1/_{216} = 10.5$$

Similarly, for variance:

V(X) 'sum of 3d6' = 3 * V(X) '1d6 ' = 2.92 + 2.92 + 2.92

$$= 3 * 2.92 = 8.75, \text{ SD(X)} = \sqrt{8.75} = 2.96$$

(cannot sum SDs of single dice, see below)

The formulae for single dice parameters (Section 18.2.1) can be augmented to give those for dice sums:

$$E(X)_{multiple\ dice\ sum} = n * (s+1)/2$$

$$V(X)_{multiple\ dice\ sum} = n* (s^2 -1)/ 12$$

In Excel: $= n * (s+1)/2$

$$= n* (s\wedge2 -1)/12$$

Where, X = sum, n = no. dice and s = dice size **(19.1a)**

[for two dice these reduce to $(s+1)$ and $(s^2-1)/6$, respectively]

Thus, for the sum of 3d6,

$E(X) = 3*(6+1)/2= 10.5$

$V(X) = 3*(6^2- 1)/12 = 105/12= 8.75$, SD $= 2.96 \approx3$

$E(X)$ '2d20' $= 2 * (20+1)/2 = 21$ (mode at 21, 5%)

The expected value of a **discrete triangular distribution** can also be calculated when the lower and upper bounds of the sum ('a' and 'b', respectively) and the mode (the most frequently appearing sum, 'c') are identified. For 2d6, these are 2, 12 and 7 respectively, thus:

Mean: $E(X)_{triangular} = (a + c + b)/ 3$

In Excel: $= (a + c\_1 + b)/ 3$ **(19.1b)**

Hence, $E(X)_{triangular}$ '2d6' $= (2 + 7 +12) /3 = 21/3 = 7$

This result agrees with (19.1a), $(6+1) = 7$, in fact the formulae are equivalent. It can also be used for the bimodal dice sum (odd numbers of dice), if the averaged mode is used. Thus, with 3 dice,

$E(X) = (3 + 18 + 10.5)/3 = 10.5 = 3*(6+1)/2$

Additionally, as dice sum distributions are symmetric, the mode equals the expected value and expression (19.1b) reduces to $E(X) = (a+b)/2$,

2d6 $= (2+12)/2 = 7$ 3d6 $= (3 +18)/2 =10.5$

With other dice,

$$E(X) \text{ '2d8'} = (2 + 9 + 16)/3 = 9 = (8+1) \text{ (mode at 9, 12.5\%)}$$

$$\text{Mean of 2d5} = (2+ 6 +10)/ 3 = 6 = (5+1) \text{ (mode at 6, 20\%)}$$

Taking the 10d6 case:

$$E(X) \text{ '10d6'} = 10 * (6+1)/2 = 35.0 = 10 * E(X)_{1d6} = 10 * 3.5 = 35$$

$$V(X) \text{ '10d6'} = 10* (6^2 -1)/ 12 = 350/12 = 29.17 \approx 10 * 2.92 = 29.2$$

$$SD(X) = \sqrt{V(X)}_{10d6} = \sqrt{29.2} = 5.4 = \sqrt{10} * \sqrt{V(X)}_{1d6} = 3.16 * 1.71 = 5.4$$

Mixed dice sizes can be accommodated, e.g. the sum of 1d8, 1d10 and 1d12 on average would give:

$$E(X) \text{'1d8 +1d10 +1d12'} = E(X) \text{ '1d8'} + E(X) \text{ '1d10'} + E(X) \text{ '1d12'}$$

$$= (8+1)/2 + (10+1)/2 +(12+1)/2 = 4.5 + 5.5 + 6.5 = 16.5$$

For the variance of a mixed dice sum, each die must be calculated using (18.2b) then added,

$$V(X) \text{'1d8 +1d10 +1d12'} = (8^2 -1)/12 + (10^2 -1)/12 + (12^2-1)/12$$

$$= 25.416667, \quad SD = 5.0414945 \approx 5$$

(To get the SD, V(X)s must be calculated separately, then added and followed by the square root of the sum)

Thus, we can easily calculate expected sums and variance for any number and size of dice.

Mode

For (19.1b) and some other calculations, the **sum mode** is required. Distributions of sum outcomes from even numbers of dice have a perfectly symmetrical appearance and the mean (expected value) and the mode are equal, lying at the centre, i.e. the expected value is equivalent to the mode. With odd numbers of dice there will be two modes and with mixed sizes three or more modes are possible (e.g. Figures above & Fig. 14.3, Book

One). The mode(s) is the most frequent sum(s) and has the highest probability. It can be presented in variety of calculation forms,

$$Mode_{dice\ sum} = (max-min)/2 + min$$

This expression can be converted to a form using the dice size and number of which there are a number, perhaps the simplest being,

$$Mode_{dice\ sum} = (s*n +n)/2$$

Where, s, n= size and no. of dice, respectively

min = no. of dice (n)

max = s*n   *(19.1.c)*

(Assuming standard numbering on faces)

With odd numbers of dice, these calculations will give a non-integer value indicating a bimodal condition.

Thus, we can see that the mean and the mode of the triangular distribution are equal. So, for 2d30:

Mean = (2+ 31 + 60)/3 = 31   Mode = 60/2 +1 = 31 (P = 3%)

For more dice,

4d6, n = 4, s = 6, mode = ((6+1) * 4)/2 = 7*4/2 = 14 (11%)

3d6 = ((6+1) * 3)/2 = 10.5 i.e. bimodal at '10' and '11' (12.5%)

These formulas simplify more if the dice size is entered:

For d6 sums: 7*n/2

For d8 sums: 9*n/2

For d30 sums: 31*n/2

... etc.

Thus, for 8d8 the most probable sum is,

9*8 /2 =72/2= 36 (6%)

4d30 =31*4/2 = 124/2 = 62 (2%)

5d20 = 5* 21 /2= 105/2=52.5, bimodal at 52, 53 (3%)

We can quickly get the most probable outcome in games with dice throws, so in a game of *Craps* and in several board games, the most likely sum outcome with 2d6 is '7'(17%). In *Dice Wars* 5d6 rolls have one of two possible mode sums at '17' or '18' (10%) and 4d6 has one at '14' (11%).

More on expected values appears later (Section 19.3.3).

The probability function

With two dice, a ***probability function*** can be formulated for two cases based on the triangular distribution, one where the sum is above the mode and the other where the sum is equal or less than it:

$$P(=k) = (k-1)/s^2 \text{ for } k<=c, \text{ and}$$
$$P = (b +1 - k)/s^2 \text{ for } k>c$$

Where, k = sum, c = mode, b = upper bound (max)   ***(19.1c)***

Thus for 2d6 and a sum of '5', which is less than the mode (=7):

$$P(2d6, sum=5) = (5-1)/6^2 = {}^4/_{36}$$
$$P(2d6, sum = 8 \text{ (above the mode)}) = (12+1-8)/6^2 = {}^5/_{36}$$

For 2d8, mode = 16/2 +1 = 9 and,

$$P(sum = 14) = (16+1 -14)/8^2 = {}^3/_{64}$$
$$P(sum = 6) = (6-1)/8^2 = {}^5/_{64}$$

It must be pointed out that these formulae appear cumbersome when, with two dice, probabilities can be worked out reasonably easily by simple counting, and the expected value and variance can be calculated by the simple procedures above. These arguments lessen the applicability of the discrete triangular distribution in dice sum probabilities, but the formulae are useful for the two-dice case with less common sizes, i.e. 2d4, 2d8, 2d10, etc. For two dice of different size, the triangular distribution parameters do not apply in the same way due to the prevalence of the multi-mode distributional form. We can calculate probability of mixed dice sums as described later.

It is not possible to apply the above formulae to the sum of three or more dice. In fact, there is no specific distribution that fits the '3 or more dice sum' random variable. Probability assessments for these sums require different approaches, some described above. Another is apparent however, as these dice distributions do approximate the *normal distribution* well, especially as the numbers of dice increase. Expected value and variance are readily available as described above and this allows calculation of probabilities by 'approximation to normal', which follows presently.

19.3 Dice sum probability and enumeration of the sequence count

For any dice, assuming standard numbering, two particular sums can be obtained by addition of their own unique sequence, but in the majority of cases, the sum can be attained by two or more different sequences. Thus, with 3d6 the sums '3' and '18' come from the sequences '1,1,1' and '6,6,6', respectively, but 27 different sequences add up to '11' and 6d8 have 10,752 sequences that give the total '21'. *'Getting a sum of 28'* on 3d10 describes a summed dice event, where any content of three face values that add-up to '28' would qualify. The probability of a sum is given by the classical probability formula (3.1) (Book One) modified for dice:

Dice Sum Probability =

$$\frac{\text{no. sequences that give the sum}}{\text{no. all possible sequences}}$$

The denominator in this expression is the count of all the sequences possible with the particular number of dice. This latter number can be calculated as described in Section 18.3.1, being essentially the product of the dice sizes. The former (numerator) count is more difficult to access, depending on the numbers of dice and availability of computing power.

Manual counting of sequences from matrix diagrams (Fig. 19.4) is simple but more onerous beyond two dice. The multi-modal nature of the mixed dice sum is displayed in the diagram. A more complex example shows more of the ability of the diagram with a combined event, which includes use of the AND operative. The example only counts sums that have BOTH properties, i.e. 'even' AND '8 or more'. This avoids possible errors of counting that can occur with visual counting from tables, e.g. 'there are more even sums but they do not sum to 8 or more'.

Lists and using a spreadsheet to generate sequences and then counting as described in Section 18.3.1 can be implemented for sums, including those for mixed dice. Referring back to Figure 18.4, we can see that it has been embellished to show sums and their counts. It shows all 216 sequences and gives a frequency count of occurrence for each sum. For this diagram at least, we can gauge the count for 3d6 sums, e.g. a sum of 8 is obtained by 21 sequences. Thus, the probability for a sum of 8 with 3d6 is $^{21}/_{216}$, a sum of 14 can be attained by 15 sequences, and P(sum=14, 3d6) = $^{15}/_{216}$, etc.

Evaluations of sum probability are also possible by a relative frequency experiment, directly or by computer simulation (Section 14.3.1). It must be emphasised that this method, while simple in terms of understanding, gives an *estimate* of frequency of occurrence of groups of outcomes for each sum.

As dice numbers increase to three or more, manual procedures can get tedious and error prone. With 3d6, there are 16 summed totals, but these are built up from 216 outcomes, and these numbers increase dramatically for the sequence count, e.g. 6d6 has 29 sums and 46656. Even generating all the sequences for larger numbers of dice as described above, would run into problems in basic Excel and construction of a macro (mentioned in Section 18.3.1) would be useful. However, it is possible to get accurate counts at these levels and beyond, with a neater 'one formula' method - but at the expense of more mathematical knowledge in *combinatorics*.

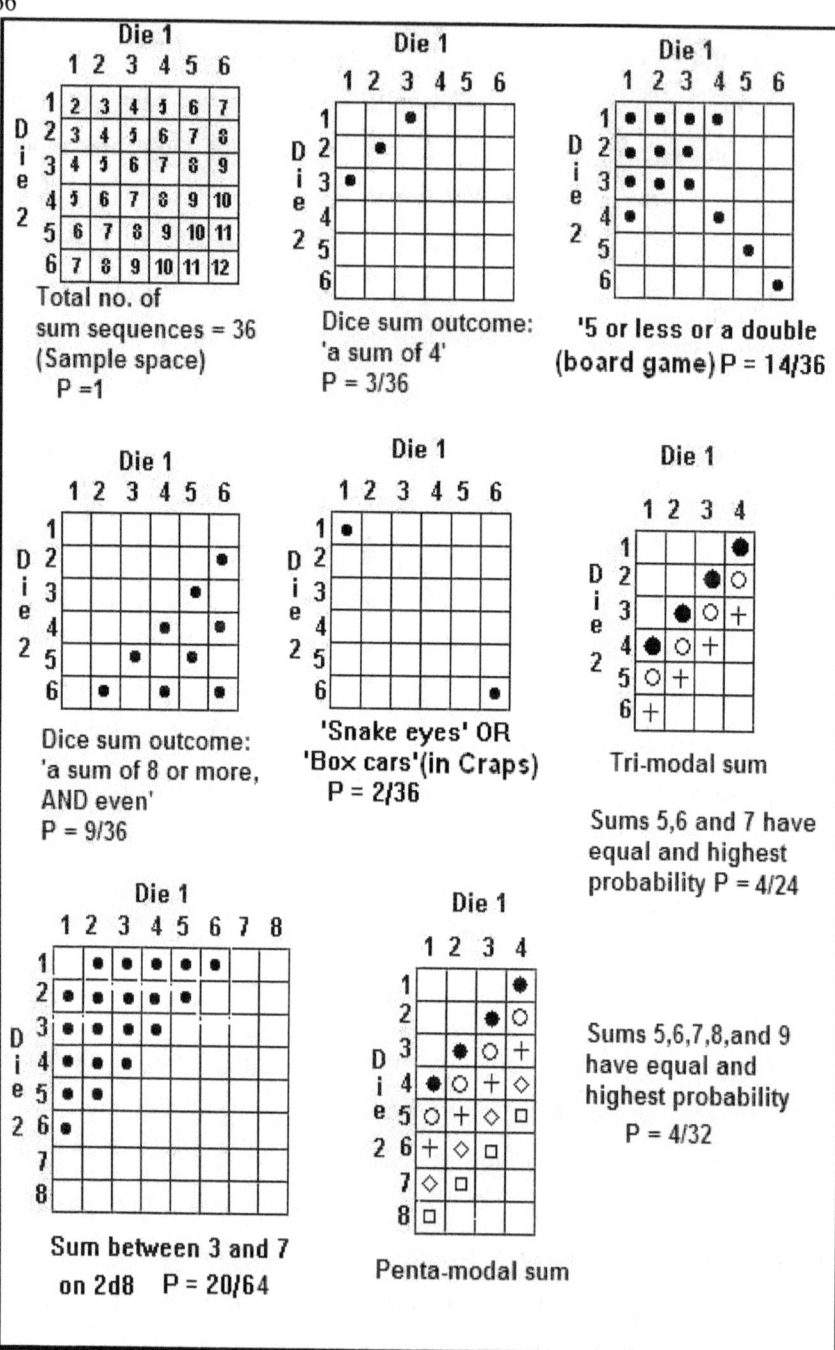

Figure 19.4 Two-way matrix diagrams for dice sum probability

19.3.1 Combinatorics for dice sums

Application of combinatorics to dice sums essentially becomes a matter of solving an equation of the possible numbers that add up to a particular sum. Taking the simplest case with two dice, 2d6 and a sum of 7, we can see that the equation,

die 1 + die 2 = 7 will have 6 solutions as,

1+6, 6+1, 2+5, 5+2, 3+4 and 4+3

To facilitate this for more dice, which will have many more solutions, combinational formulae are used to count all possible, non-zero integer solutions, based on formula (7.5b) (Book One) and as detailed in McShane and Ratliff. Subsequent steps involve adjusting an initial solution for any results that exceed the dice size, and subtraction of any repeated counts. The expression is evaluated one or more times according to the sum, number and size of the dice, with a repeating multiplier of '+1, -1, +1 ...' signifying the 'inclusion-exclusion' effect. Thus, in the case of two dice the expression is evaluated once or twice depending on the magnitude of the sum, whereas with 20d6 and a maximum sum, 19 evaluations are required. The evaluations are then summed to give the final count, which is multiplied by the reciprocal of the total number of outcomes to calculate the probability. The formula is stated in several sources and the following is adapted from that of Epstein (p. 176):

Formula for dice sum probability:

$$P(k) = s^{-n} * SUM[(-1)^{j} * {}^{n}C_{j} * {}^{k-s^{*}j-1}C_{n-1}]$$

from 0 to j

In Excel: see below

Where, n, s = no. & size of dice, k = sum, j = INT[(k-n)/s] *(19.2a)*

So, for 2d6 and a sum of '7', n = 2, s = 6, k = 7,

$$j = INT[(7-2)/6] = INT(5/6) = 0$$

The expression is evaluated according to the value of j, starting at 0 - so in this case, <u>once</u>:

$$P(7) = 6^{-2} * [(-1)^0 * {}^2C_0 * {}^{7-6*0-1}C_{2-1}] = \frac{1}{36} * 1 * 1 * {}^6C_1 = \frac{6}{36}$$

For a sum of '10', $j = INT((10-2)/6) = INT(8/6) = 1$ and the expression is evaluated <u>twice</u>, i.e. at $j = 0$ and $j = 1$:

$$P(10) = 6^{-2} * [(-1)^0 * {}^2C_0 * {}^{10-6*0-1}C_{2-1}] + [(-1)^1 * {}^2C_1 * {}^{10-6*1-1}C_{2-1}]$$

$$= \frac{1}{36} * [1 * 1 * {}^9C_1] + [-1 * 2 * {}^3C_1] = \frac{1}{36} * (9 - 6) = \frac{3}{36}$$

With low values of 'j', the expression reduces to simpler forms but calculation of more elaborate cases is best left to a computer (see below). There are other, more complex ways to calculate these probabilities including 'decomposing' complex expressions to produce simpler base values that can be used recursively to sum the specific ways, or to use generating functions.

To implement the above formula (19.2a) in Excel, the various data values are named (**Define Name**) first (note, Excel does not permit use of [ ] brackets):

n = no. dice , s = dice size, k = sum,

j = maximum no. of evaluations (set = INT((k-n)/s)

The main expression is entered as,

(-1)^j)*COMBIN(n,j) * COMBIN((k-(j*s))-1,n-1)

Where, j varies from 0 to maximum j

This must be evaluated for a total of j+1 times then totalled using **SUM( )** and multiplied by the reciprocal of s^n:

P(sum = k, on n dice of size s) = **s^-n * SUM(...)**

Thus, for 3d6 and a sum of '16', $n = 3$, $s = 6$, $k = 16$ and $j = 2$. The expression is evaluated 3 times, with the multiplier varying as +1, -1, +1 and j from 0 to 2:

Spreadsheet table 19.1

	B	C	FORMULA
	j=		
10	0	105	= (-1)^B10 *COMBIN(n,B10) * COMBIN((k-(B10*s))-1,n-1)
11	1	-108	= (-1)^B11 *COMBIN(n,B11) * COMBIN((k-(B11*s))-1,n-1)
12	2	9	= (-1)^B12 *COMBIN(n,B12) * COMBIN((k-(B12*s))-1,n-1)
13		6	= SUM(C10:C12)
		0.027778	= 1/s^n *C13
		1/36	= P as fraction

Total no. sequences = 105-108 +9 = 6 and $P = 6/6^3 = {}^1/_{36}$ (3%).

(With lower sums and the adjustments explained below, this reduces to $P = {}^4C_2/6^3$).

The number of evaluations required depends on the data, with higher sums for larger nos. of dice generating more repeats requiring more adjustment. Excel can run into precision errors with very large numbers and converting all target sums to the lower half of the dice distribution (Section 19.2) simplifies the calculation. The probabilities are the same because of the symmetry. For example, 17d6 with a sum of 78 would require 10 evaluations but the corresponding lower sum (41) needs only 5. To get the 'mid-point', use the **mode** (**m**) of the distribution (19.1c), which for d6 sums, is given by 7*n/2 and,

mode = 119/2 = 59.5

Provided the original sum k > (mode +.5), then,

Lower sum required = 2 * m - k   *(19.2b)*

Essentially, the amount above the mid-point is subtracted from the mid-point.

Hence, for sum (17d6) = 78, lower sum (i.e. new k) = 2 * 59.5 – 78 = 41

Using Excel for this sum,

$$j = \text{INT}((41-17)/6) = 4,$$

so the expressions is evaluated 5 times, giving the output:

Spreadsheet Table 19.2

j =	
0	62852101650
1	-37467344310
2	4137358680
3	-50736840
4	2380
No. of specific ways =	29471381560
No. of total ways =	16926659444736
P(17D6, sum = 78) =	0.001741122142631
fraction =	1/574
% =	0.17%
Odds against to 1	573
Odds in favour to 1	0.001744159

Thus, $P('17d6\ sum=78') = 0.00174112\ (0.2\%)$

Additionally, calculations are simpler when the numbers of dice and size with the sum result in 'j' being calculated as 0 or 1, e.g. for 10d6 and sum = 15, $j = \text{INT}((15-10)/6) = 0$ and the count of qualifying sequences reduces to $^{k-1}C_{n-1}$:

$$P('10d6,\ sum=15') = \frac{1}{6}^{10} * {}^{10}C_0 * {}^{15-6*0-1}C_{10-1}$$
$$= \frac{1}{6}^{10} * {}^{14}C_9 = \frac{2002}{60466176} = 0.0000331094$$

$$P('7d12,\ sum=81;\ lower\ sum\ 10')$$
$$= \frac{1}{12}^{7} * {}^{10-1}C_{7-1} = \frac{84}{35831808} = 0.00000234429$$

In Excel worksheets, formula (19.2a) is accurate up to 17d6. Beyond this some numbers exceed the 15 digit limit[5] (Excel 2003 -2010) and results will

---

[5] Maximum no. with 15 digits is 1000 billion minus 1 (999,999,999,999,999)

become more and more approximate, although relative errors are extremely low (-0.00000000000002% for 20d6 on sum of 70, -0.04% on 50d6 for sum of 175). Excel (2010) can store much larger numbers but only to 15 digit accuracy (Walkenbach).

Eventually, above 50d6 or the equivalent in other dice sizes and numbers, relative error increases and the counts of total or specific ways may go negative. At this point, an estimate can still be had by a normal approximation (see below). The expression can be used with other dice sizes (but not with mixed sizes), but larger numbers of dice faces reduce the number of dice for full accuracy.

Any sum 'at least' and 'at most'

The probabilities for these events requires adding together (the addition rule (OR')) the individual probabilities for each sum in the requirement. For example, *'getting a sum of at least 11 with 4d6'* requires summing the initial exact sum probability and those <u>above</u>:

$P('sum at least 11, 4d6')$

$\qquad = P('4d6, sum=11') + P('4d6, sum=12') + P('4d6, sum=13')$

$\qquad + P('4d6, sum=14') + ... P('4d6, sum=24')$

$\qquad = {}^{1090}/_{1296} = 0.841048 \ (84\%)$

Correspondingly, for 'at most' sums, summing the probabilities of the initial exact sum and those <u>below</u> is necessary. Watch for the facility of using the complementary event, e.g. *'4d6 at most 20'* – instead of summing '4 to 20' use $(1 - \text{sums '21 to 24')}$.

Thus, *'getting 10 or more with 2d6'*

$P = P('2d6, sum =10') + P('2d6, sum =11') + P('2d6, sum = 12')$

$\qquad = 0.083 + 0.056 + 0.028 = 0.166667$

$\qquad = {}^{3}/_{36} + {}^{2}/_{36} + {}^{1}/_{36} = {}^{6}/_{36} = {}^{1}/_{6} = 0.166667 \ (17\%)$

'At most 10' would be the complimentary event of *'11 or more'*,

$$P = 1 - {}^3/_{36} = {}^{33}/_6 = 0.916667 \ (92\%).$$

As seen, using the fractional form makes for easy arithmetic. When the above procedure gets laborious with larger ranges of sums or your interest lies in exploring sum probabilities beyond 50d6 or equivalent, then an alternative route is to employ a normal approximation (Section 19.3.2), which can calculate sum probability in an 'at least' and 'at most' manner.

Alternatively, you can set up a spreadsheet with formula (19.2a) to sum the counts for several values above or below any input. All sums for the particular number and size of dice must be put in a row with the formula (Spreadsheet Table 19.1) copied in below each sum to cover equal or more evaluations required with the count summed at the foot. Two new variables are named and entered below the summation: 'from_sum' and 'to_sum'. In the example, a row would be created with sum data values 4 to 24 and the variables would be equal to 11 and 24 respectively. Then, if the row goes from H42 to AB42, the count is calculated with the following, entered below the two variable inputs:

= SUM(INDEX(H42:AB42, from_sum - n+1)
  :INDEX(H42:AB42, to_sum - n+1)) = 1090,

confirming as above (a longer row can be set up to cover more sums).

19.3.2 Normal approximation for dice sums

Based on the central limit theorem, the distribution of a sum of independent random variables is approximately distributed as normal with mean = n * E(X) and variance = n * V(X). These parameters are calculated as above (Section 19.2.1). Excel functions are used for the approximation. Any required range of sums can then be calculated, plus an estimate of an exact value location.

Probability function for dice sum approximation

The Excel NORM.DIST( ) function (Book One Section 12.2) is used for these problems, with the cumulative argument set to 'TRUE', which integrates all values within a specified range. The sum can be one value ('k') or a range spanning a lower and upper bound ('lk' and 'uk', respectively), where 'lk' is equivalent to the lower sum (can go as low as the minimum sum ('n')) and similarly for the upper sum 'uk', (up to the maximum sum (s*n)). The probability is calculated essentially as described in Book One (Section 12.2) and in Section 15.2.2. The average sum and the standard deviation are required (via the E(X) and V(X) for single dice), then:

P('sum up to k') = NORM.DIST(k+0.5,ave,SD,1)

P('sum k or more') = 1 - NORM.DIST(k-0.5,ave,SD,1)

P ('sum between lk and uk') =

= NORM.DIST(uk+0.5,ave,SD,1) – NORM.DIST(lk-0.5,ave,SD,1)

Where, ave = $E(X)_{multiple\ dice}$ SD = $\sqrt{V(X)}_{multiple\ dice}$

k, lk = sum, lower bound,   k, uk = sum, upper bound   **(19.2c)**

For discrete distributions, a continuity correction ($\pm$ 0.5) is required to counteract slight inaccuracies caused by using a continuous probability model for the discrete situation. Additionally the approximation improves with the similarity of the dice discrete distribution to the normal curve.

We can now calculate *'between'*, *'at least'* and *'at most'* sums. Using a 10d6 example ('ave' = 35, 'SD' = 5.4):

P(*'10d6, up to a sum of 28'*) = NORM.DIST(28+0.5,35,5.4,1)

= 0.114352 (11%) [accurate value of P(*'10 to 28'*) = 0.116047]

(Note: Calculating P(*'a sum between 10 and 28'*)

= NORM.DIST(28.5,35,5.4,1) – NORM.DIST(9.5,35,5.4,1)

is virtually the same at 0.114351)

P('*10d6, greater than 28'*)

= 1– NORM.DIST(30-1-0.5,35, 5.4,1) = 0.885648 (89%)

[P('*29 to 60'*) = 0.883953]

P('*10d6, a sum between 20 and 40'*)

= NORM.DIST(40+0.5,35,5.4,1) – NORM.DIST(20-0.5,35,5.4,1)

= 0.843734 (84%) [P('*20 to 40'*) = 0.842016]

These approximations appear to have excellent correspondence with the accurate results determined as above (combinatorics, Section 19.3.1). We can now calculate estimates for larger numbers of dice,

'*74d6, sum of between 100 and 250* ',

ave = 259, SD = 14.7, lk = 99.5, uk = 250.5,

P = 0.281554 (28%)

Approximations for exact dice sums

As explained in Book One (Section 12.2), the normal distribution function, being for a continuous random variable, cannot calculate probabilities for exact discrete values. However, the above treatment ($\pm$ 0.5 around a sum 'k') can be applied to an <u>exact</u> sum to give a small range (although Excel does provide a probability mass setting). Thus:

'*Sum of 69 on 20d6*'   ave = 70, SD = 7.64, k= 69, lk = 68.5, uk = 69.5

= NORM.DIST(69.5,70,7.64,1) – NORM.DIST(68.5,70,7.64,1)

P = 0.0517359 (5%)

The mass function (cumulative = 0), gives,

P = NORM.DIST(69,70,7.64,0) = 0.0517722

The accurate value (by combinatorics) is 0.0514, so both methods have produced very close approximations (rel. error $\approx$ 1%; formula (14.1), Book One Section 14.2.1) with a slight over estimate. Another example shows an underestimate,

$$P(10d6=28) = NORM.DIST(28,35, 5.4,0) = 0.0318879$$

(agrees with the other calculation)

The exact value is 0.0326327, producing rel. errors of approx. -2%, higher than above but the confidence interval (CI; Book One Section 14.2.1) covers the range 27.5 to 28.5.

For very large numbers of dice, '*Sum of 370 on 100d6,*

ave = 350, SD = 17.08, k= 370, lk = 369.5, uk = 370.5

$$P = 0.0117680 \ (1\%) \ (\text{accurate value} = 0.0117878; \text{rel. error } -0.17\%)$$

The accuracy of the dice sum approximation varies, generally increasing with the numbers of dice. Best agreement between exact probabilities (if available) and the approximations are found with dice sums located within ± 2*SD, getting progressively less so near the tails of the distribution (see Problems & Exercises).

19.3.3 Expected value of dice sums

The distribution and the properties of dice sums were covered above (Section 19.2.1), where the $E(X)$ of a sum is given by the addition of the single dice $E(X)$. Having dealt with the sum probability calculation, we can extend this to some other events.

The expected appearance of a particular sum in any number of <u>repeated trials</u> can be calculated using the binomial $E(X)$, namely, n*p, where, for this application, n = no. of repetitions of the original experiment and p = probability of a particular event within the original (Book One Section 8.2.2). For example, in a fantasy combat a fire projectile does 10d6 damage – during a typical game a magic arts fire wielder uses the projectile about 5 times, so how often would a sum of 60 points appear?

$$P \ ('sum = 60, \ 10d6') = \frac{1}{6}^{10}, \text{ thus } n = 5 \text{ and } p = \frac{1}{6}^{10}$$

$$E(X) \text{ 'on 5 uses'} = \frac{5}{6}^{10} \text{ (a very low incidence)}$$

The average sum for 10d6 is 10 * 3.5 = 35 points so maximum damage (60) rarely appears. Getting a lower amount, say '40 or more' with probability at 20%, ('sum of 40 or more' using (19.2c)), should occur on average once during a typical game as $E(X) = 5 * \frac{1}{5} = 1$.

Another aspect is the **expected number of throws** to get a particular sum, e.g. 'with 2d4 how many throws are required to get a sum of 7?' This calculation can be based on the geometric distribution expected value provided we know the probability for the sum. Referring to Table 18.2, the probability P('sum = 7, 2d4') = $\frac{1}{8}$. The expected number of trials (throws of 2d4),

$$E(X) = 1/p = 1/ (\tfrac{1}{8}) = 8$$

Similar calculations can be applied for more than one success with the corresponding negative binomial parameter, such as 'how long will it be before I convert this point? (Craps) or 'How long do I need to roll before I get a decent total in this game? For these and others, see next chapter.

Chapter 20
Dice games and dice problems

20.1 Introduction

There are many games where dice are used. Like other games, some accounts of the game structure take the mathematical level beyond this text, but there are instances where we can use the methods of the previous two chapters to highlight the application of probability, remembering the points discussed in Section 15.2.

In *dice games* themselves, face values and sums are used in ranked order or in scoring combinations ('patterns') to establish a winner or a maximum. The function of dice in other games often focuses on a method for deciding actions, e.g. in many *board games,* dice decide the movement jumps round the board and some other functions. Choices may be available as the die or dice are rolled, and rolling a designated number(s) or sum sometimes allows a re-roll or further actions.

In *war games,* dice are used mainly in deciding combat outcomes. In *role-play games*, combat actions also feature and in some cases, polyhedral dice are used. Another feature is the use of dice for determining the characteristics of the persona in such games.

To summarise, in games using dice as the main or only component, the randomising device(s) can be used in various ways, classified roughly as:

- Getting an exact match or inequality with a target number on a single die, e.g. a '20' on 1d20 in role-play combat
- Getting certain combination of *face values* as sequences or content on 2 or more dice, e.g. '4 of a kind' in *Poker dice*
- Getting an exact total or an inequality with the *sum* of 2 or more dice, e.g. a '7' in *Craps*

The accounts below are based roughly on these divisions, but there are many instances of overlap. Nowadays, many dice games and games where dice are used are available in video game format on PCs and tablets. The probability for these games is sometimes provided, which can allow you to compare with those given below. In other cases, probability is hidden and may be adjusted for bonus 'boosts' and penalties, etc. but this is unknown to the user.

20.2 Probability and games using dice face values

Here the game structure is based on scoring points or controlling actions according to the appearance of one or more *dice face values*, usually as a count, e.g. 'three 4s' wins this turn. Some have choices at stages in the game, where probability can be used to guide the way. With one die, the magnitude of the face is of importance as it decides the issue, e.g. *'you need to roll a 1 to get out of this room'* or *'I've rolled a 9 on the d12 so I can move over the obstacle'*.

We start with some examples of the single die case, more of which appear in Section 20.3.

Board games

With many *board games* and similar games a '6' (or a '1' in some) on 1d6 is required to start. Your chance of winning at this stage is explained above (Section 15.2.1). All players have an equal chance, but ties are possible, although a set of 'one roll dice' have been devised (www.ericharshbarger.org) to avoid the possible tedium of having to do re-rolls. Single die probabilities are straightforward and are calculated with formula (18.1). Traditional favourites like *Snakes and Ladders* and *Ludo* use a single d6 to monitor movement, so for any move, base probabilities are $^1/_6$:

'I need a '5' to get to the bottom of the ladder' - $P = {}^1/_6$

'Watch that snakehead 2 squares in front - roll to avoid it' -

$P('1,3,4,5,6') = {}^5/_6$ but ${}^4/_6$ ($'3,4,5,6'$) to get beyond.

'Ludo - If I get a '3' I can finish this piece' - $P = {}^1/_6$

'Ludo - A '4' will let me bump you off the board' - $P = {}^1/_6$

Some of such games have rules that say that you must roll the exact number to finish - e.g. you are 4 away – how long will it take you to get a '4'? This is the expected value of the geometric distribution $(1/p) = 6$, so six times on average, during which your opponent may be hot on your heels! Assume the opponent is Lulu and she is 8 squares behind – you fail to get the '4' and get '5' or '6' (${}^2/_6$) so you stay where you are– what is probability of her overtaking and winning? If she gets a '4' (${}^1/_6$) she 'bumps' you back to the start, any higher or lower (${}^5/_6$) then she either takes the lead or remains behind you. This opens up several pathways for the next moves, some with complex probabilities with all the 'bump', 'finish' and 'overtake' possibilities.

Also in *Ludo*, a choice to start another piece or to move a piece already in play occurs when a '6' is rolled. The action taken depends on how many pieces you have in play and on the position of opponent pieces – if they are close behind, then, if the particular version of the rules allows it, moving 6 squares will widen the gap and getting an active piece into safety may be the wiser game choice. Otherwise, getting a sleeping piece active is the choice rather than have to wait and lose opportunities when the time comes, based on an average of six rolls to get a '6'. A **run** (Section 15.2.3) of '6s' although infrequent, can turn the game around (some rules say a 3rd '6' loses the turn, others don't - so probabilities go ${}^1/_{36}$, ${}^1/_{216}$, etc. On average it takes approx. 20 rolls per piece to get home (70 spaces divided by the E(X) for 1d6) giving around 100 per game when mishaps like 'bumps' are built in. With this number of rolls, there is a very good chance (approx. 90%) of 1 or more runs of two '6s' using the POISSON( ) function as demonstrated in Section 15.2.3 (32% for three '6s').

Single dice in other games

In other games, the single die outcome can decide success or fail of any action and there are examples with all sizes of dice, covering a miscellany of actions and effects in tabletop games, adventure and role-play games, such as cost of items purchased, no. of prizes won, penalty levels, sales levels, avoiding traps, jumping, climbing, lifting and of course, engaging in battles and melee. (In war games and role-play where there is a combat element employing a single die, the outcome can also be viewed as a 'sum' and more is said on this in the next section). Some examples in various board games are:

'Your movement / the number of actions is decided by the roll of the dice'

'You roll a die and multiply by the prize money/ penalty'

*'If you roll a certain value or less you suffer a disadvantage (*typically miss a turn*) / gain an advantage* (e.g. roll again)

Single die vs. single die can be used in any contest between protagonists in games giving the duel situation (Section 15.2.3). *'I can jump further than you!'* says one adventurer - 'Ok *prove it'* says another. Each rolls the appropriate die for the system and the higher wins. Different sizes of dice can be used in these situations to represent differing levels of ability.

Two dice face values

For two dice, the count of possible face values increases (0,1,2). Often, a particular face becomes the objective for the roll and the ***binomial function*** can be used to get the probabilities for the occurrences. Two repeated face values often have a special significance in board games as with 'rolling a double' on 2d6. Like the single die case, success allows rerolls (up to three times is some games but more in others). We know from previous calculations that the probability for a double on 2d6 is $^6/_{36}$. We treat this as a single compound event, then,

P(*'double once on 2d6'*) = BINOM.DIST(1,1,6/36,0) = 0.166667 (17%)

Using the function again gives *'twice in a row'* (P = 0.0278 (3%)) and *'thrice'* (P = 0.00463 (0.5%)). Thus, three doubles one after the other is a rare event, but some games have a penalty on the third double (see the **run** above). Some war games use one or more dice where a particular face is counted during the combat stage (see below) but now we look at some pure dice games.

20.2.1 Dice configuration pattern games

In this section, we consider *dice games* as such with the dice being the focus. In these games, dice are thrown and winnings in the form of monies or points are awarded based on the attainment of various 'patterns' of configuration. At the simplest level, this can be the occurrence of one or more of a particular face on the dice. Games that are more elaborate have more dice allowing more styles of scoring patterns, such as repeated face values and consecutive numerical sequences.

Three dice

An example of the former type of dice game is found with the three-dice games *Chuck a Luck* and a similar game, *Crown and Anchor*. For the latter, the dice can be marked with symbols ('crown', 'anchor' and the four card suits), but ordinary d6s can be used. Bets are placed on the occurrence of one or more face values. Probabilities can be calculated via the 'unnamed' dice expressions (Section 18.3.4). In the event, $'\Psi,x,y'$ ('anchor,x,y', assumes one anchor on dice), we can get distinct and repeated in the 'x,y' part, i.e. any outcome with a single anchor qualifies. Using the specific case method (18.9c), we have two possibilities,

Specific case 1, distinct unnamed $\rightarrow$ $'\Psi,\text{♛},\text{♠}'$ (anchor,crown,spade)

$$P = \frac{1}{6}^3 * {}^5C_2 * 3! = \frac{60}{216}$$

Specific case 2, repeated unnamed $\rightarrow$ $'\Psi,\text{♦},\text{♦}'$ (anchor,diamond,diamond)

$$P = \frac{1}{6}^3 * {}^5C_1 * 3!/2! = \frac{15}{216}$$

These are added (the addition rule) = $^{75}/_{216}$

$$= 0.347222 \ (35\%, \ 2 \ to \ 1 \ against)$$

Any 2 of a face,

Specific case, distinct unnamed $\longrightarrow$ '♛,♛,♣' (crown,crown,club)

$$P = \frac{1}{6}^3 * {}^5C_1 * 3!/2! = {}^{15}/_{216} = 0.069444 \ (7\%, \ 13 \ to \ 1 \ against)$$

All 3 of a face,

Specific case, no unnamed $\longrightarrow$ '♥,♥,♥'

$$P = \frac{1}{6}^3 = {}^1/_{216} = 0.00462963 \ (0.5\%, \ 215 \ to \ 1 \ against)$$

Results include the approximate (but true) odds against. We can also use the **binomial function**, to confirm the first calculation above and for an **inequality** case,

$$P('1 \ face \ 3d6') = \text{BINOM.DIST}(1,3,1/6,0) = {}^{75}/_{216} = 0.3472222$$

'At least 1 win with a single bet' (e.g. bets 1 unit on 'anchor'),

$$P = 1\text{-BINOM.DIST}(1\text{-}1,3,1/6,1)$$

$$= {}^{91}/_{216} = 0.421296 \ (42\%, \ 1 \ to \ 1 \ against)$$

This latter calculation gives more of reflection of how bets are placed in the game, i.e. placing a bet on 1 face value can win by getting 1 occurrence or 2 or 3 on the dice.

The **expected return** is related to the number of occurrences. When played in a casino or fairground, with a bet of 1 unit on 1 face, a win gets a return of 1:1; 2 occurrences get 2:1 and 3, 3:1. These undercut the true odds above, markedly. Multiple bets are also possible, e.g. betting on 2 numbers can win as – 'get 1 or get 2 or get both'. A win with 1 pays 1 and if both win 2:1 (4). All returns (ER) are negative, and play on a single outcome bet (0.42 probability) has,

$$ER = -1* 125/216 + 1 * 75/216 + 2 * 15/216 + 3 * 1/216$$

$$= (-125 + 75 + 30 + 3)/216 = -17/216 = -0.0787037$$

The idea of this type of game can be extended to more dice, giving more options, so the following game (*Poker dice*) can be played in this manner with bets on the number of occurrence of the faces 'A K Q J T 9'.

Poker dice

Perhaps the most well-known dice game for the game structure described above (i.e. getting various configurations or 'patterns' of face values), is *Poker dice*, followed by similar games like *Yahtzee®* and *Farkle*, with their many variations. The patterns are similar to those of the poker card game (pairs, trebles ('3 of a kind' (3oak)), straights, etc.) but there are significant differences for some probabilities due to the dice being limited to 6 ranks and no suits. The games can be played with dice on the table top, but many are available (free in some cases) as video versions on computers and tablets. Standard d6s can be used but poker dice are labelled with 'A, K, Q, J' pictures and '10' and '9' as pips.

More complex probability calculations are required to solve these games. This time a number of certain face values need to occur, sometimes along with unnamed outcomes. The basis of the calculations are described in Section 18.3.4. Initial probabilities on the first throw are based on 5d6 for *Poker dice* and *Yahtzee®* (*Farkle* 6d6). Face value occurrences are counted as *content* (no order considered). The constraints will usually be that the remaining dice outcomes cannot equal one another and the named ones.

To illustrate this take a 'single pair' on *poker dice*, such as 'two Ks', OR 'two 9s', OR 'two Js', etc... The calculation basis was described above (formula (18.9c)) and in Excel it can be done as,

P('*any one pair, poker dice*)

Specific case, distinct unnamed $\longrightarrow$ '*KK,x,y,z*'

$$= (1/6)^5 * \text{COMBIN}(5,3) * \text{FACT}(5)/(\text{FACT}(2)) * \text{COMBIN}(6,1)$$

$$= 3600/(6^5) = 0.462963 \ (46\%)$$

Alternatively, using the multinomial calculator (Section 18.4, Spreadsheet Table 18.3) all the patterns can be determined. We can see that the pattern is 2 0 1 1 1 0 which has $6!/3!2! = 60$ forms and each in $5!/2! = 60$ arrangements, so the count $= 60 * 60 = 3600$, as above. Similar steps are used for more matching numbers, as with three, four and five 'of a kind' ('oak'), given in Table 20.1 along with other initial probabilities, in multinomial format. Note that later rolls may assume a target, e.g. converting a small straight to a large, 1d6 is used - this differs from rolling from the beginning for a large straight.

Two pair requires an additional calculation for the number of this type possible as,

No. of 'two pair' possible $= {}^6C_2 = 15$,

if the individual probability method is used but the multinomial encompasses this. An adjustment is also required for a *full house* (3oak with 2oak). A similar initial calculation (as for two pair) gives the number possible, but this time we can have them the other way round as well (e.g. '1,1,2,2,2' and '1,1,1,2,2 etc.), thus,

No. 'full houses' possible, $= {}^6C_2 = 15 * 2 = 30$,

with $5!(3!*2!) =10$ arrangements and,

$P = (30*10)/6^5 = 0.0385802$ (4%) (but the multinomial is again simpler)

A *straight* is a pattern with 5 different numbers in a numerical sequence (Section 18.3.3). There are 2 sequences possible with the d6 (i.e. 1 1 1 1 1 0 and 0 1 1 1 1 1), each in $5! = 120$ arrangements, thus $= 2 * 5! = 240$ (some rules exclude the straight pattern but if included it is outranked by a full house, although of lower probability).

For a roll of five dice to get none of the above can be obtained by subtracting the count of all the patterns from the total,

$P = (7776 - 7296) / 6^5 = 480/6^5 = 0.0617284$ (6%)

Table 20.1 Some initial and re-roll probabilities for dice games (5 & 6d6)

	P (in 1)	Calculation
Poker dice 5d6:		
1 pair	0.462963 (46%)	$6!/(3!*2!) * 5!/2! /6^5$
2 pair	0.231481 (23%)	$6!/(3!*2!) * 5!/(2!*2!) / 6^5$
3 oak	0.154321 (15%)	$6!/(3!*2!) * 5!/3! /6^5$
5 straight	0.0308642 (3%)	$2 * 5! = 240/6^5$
'nothing'	0.0925926 (9%)	$((7776-7296)+240)/6^5$
'non-straight'	0.0617284 (6%)	$(7776-7296) /6^5$
Yahtzee 5d6:		
Small straight	0.123457 (12%)	see text
'yahtzee' (5oak)	0.000771605 (0.1%)	$6!/5! * (5!/5!) /6^5$
Treble (inc. full house)	0.192901 (19%)	Treble + full house
Farkle (6d6):		
'1 & 5' + any?	0.11574 (12%)	$(6!/2! * {}^4C_3 * {}^3C_1 + 6!/(2!*2!) * {}^4C_2* {}^2C_2)/6^6$
3 pairs	0.0385802 (4%)	$6!/(2!*2!*2!) * {}^6C_3 /6^6$
2 trebles	0.00643004 (1%)	$6!/(3!*3!) * {}^6C_2 / 6^6$
6 straight	0.0154321 (2%)	$1 * 6! / 6^6$
farkle	0.0231481(2%)	${}^4C_4 * {}^4C_2 * 6!/(2!*2!) / 6^6$
6oak	0.000128601(0.01%)	$6!/5! * 6!/6! /6^6$
Specific conversions and targets (5d6):		
Conversion	**in 1 re-roll**	**Calculation**
Pair to 3oak (re-roll 3 dice)	0.277778 (28%)	$5!/(2!*3!) * 3!/3! /6^3$
4-str. open both ends (re-roll 1 die)	0.333333 (33%)	$2/6$

Target	in 1	in 2	in 3	Overall
5 straight	0.0308642 (3%)	0.0617284 (6%)	0.0497257 (5%)	0.142318 (14%)
5oak (yahtzee)	0.000771605 (0.1%)	0.0118599 (1%)	0.0333972 (3%)	0.0460378 (5%)
Full house	0.0385802(4%)	0.207058 (20%)	0.349283 (35%)	0.588479 (59%)

Comparing these figures with those for card poker (Section 22.2.1 & Table 22.1), we can see that pairs and straights are more likely with dice and 'nothing' ($\equiv$ 'high card') is of lower probability. One consequence is that the <u>dice pattern ranks</u> for straight and full house do not match their probabilities (see below).

Improvement rolls in poker dice and similar games

These initial rolls can be improved by taking out odd dice and re-rolling. Improvement formats are similar to the 'draw' process in the card game, but with dice games, up to two re-rolls are allowed in some rules, using from 1 to 5 dice. In *Casino poker dice*, 'draw' versions allow none, 1 or 2 re-rolls, with corresponding lowering of pay outs, contrasting with a 'stud' version (no draw but higher pay outs; www.realmoneygames.org).

Probability for re-rolls can be done for each stage then for the overall chance (from 1 to 3 rolls in total if rules allow). In some cases, there are alternative ways of achieving the target (intended) pattern, e.g. 'one pair' to a 'treble' can be achieved in one way with one re-roll by keeping the pair and re-rolling three dice:

$$P(1 \text{ re-roll}) = \frac{1}{6} * \frac{5}{6} * \frac{4}{6} * {}^{3}C_1 * 1! = \frac{60}{216} = 0.278 \ (28\%)$$

But 'one pair' to 'full house' can be done in two ways, *'pair +1 +2'* or *'pair + 3',*

$$P = \frac{1}{6} * \frac{1}{6} * \frac{1}{6} * {}^{3}C_3 * 3!/3! *5 = \frac{5}{216}$$
$$\text{OR} \quad P = \frac{1}{6} * \frac{1}{6} * \frac{1}{6} * {}^{3}C_3 * 3!/2! *5 = \frac{15}{216}$$
$$\text{Net} = \frac{20}{216} = 0.0925296 \ (9\%)$$

Thus, some improvements involve many more possible ways to achieve the target when three rolls are allowed, as the extra roll means that even though the 2nd roll fails, a third is possible. These pathway variations apply from the start and it is possible to follow paths, as with getting a full house. It can be achieved outright or by converting a non-qualifying pattern (Fig. 20.1).

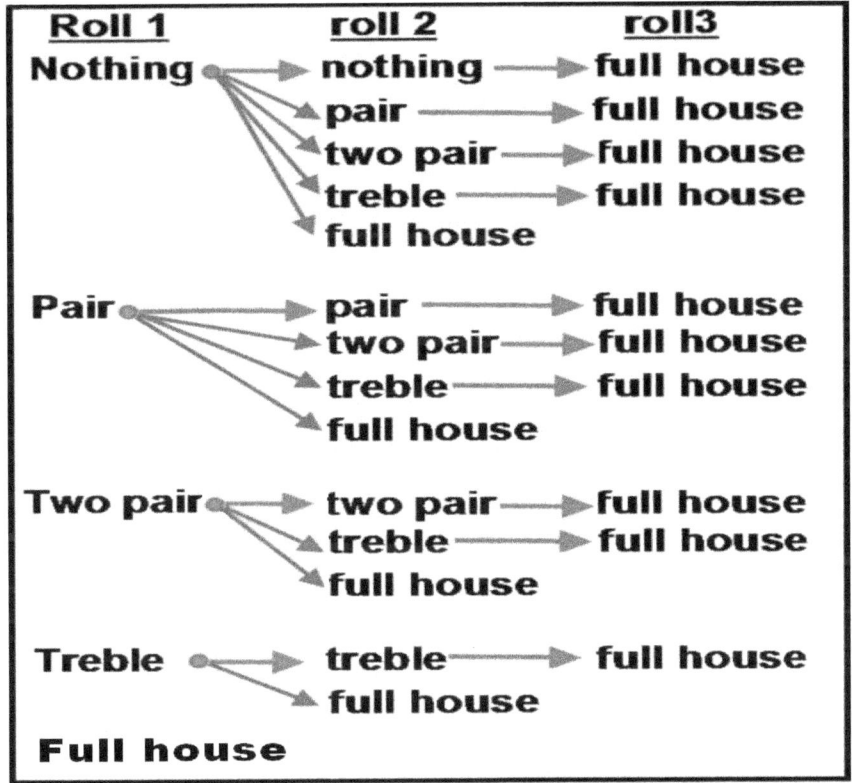

Figure 20.1 Some probability pathways in obtaining a full house in 1 to 3 rolls of 5d6

Therefore, these <u>and others</u> would be included in the complete full house probability determination. Some of these calculations assume that the end pattern is targeted - viz. if you want a full house from the start and roll a straight – this would count as nothing.

Calculated values for some other improvements are given in Table 20.1, with an abbreviated guide. Probabilities (or counts) should be summed to check the figures, e.g. all the possible outcomes on re-rolling 3 out of 5 dice from an initial pair:

Add all these and,

1+20+15+60+60+60 = 216 (= sample space) and P = 1

In cases where a choice is made on re-throws, there may be more than one scoring outcome possible and the probabilities can be compared. Provided fixed pay outs (money or points, etc.) are awarded (e.g. casino poker dice) or when there is betting with a pot, the expected return can be calculated (see below and a worked example with *Farkle*).

Poker dice probability and play

Overall strategy with *Poker dice* is, of course, to keep rolling to get the best pattern, especially so if you throw first. Patterns are ranked as the card game, viz. pair, two pair, treble, straight, full house, 4oak, 5oak (in casino poker dice, the ranks match the probability: two pair, treble, full house, straight, royal straight, 4oak, 5oak).

In play, there may be tactical choices at intermediate stages. For example, consider an initial roll that produces 'T,T,J,Q,K' - this roll has a pair and an 'open 4 straight'. Does the player go for two pair or more as above or try to convert the 4 straight? Obviously, by re-rolling three dice you get an overall chance based on the sum of those for two pair, treble, full house up to 4 oak and 5oak. The probabilities for these are worked out in a similar manner to the examples above and summed,

P(convert single pair to higher pattern on 3d6)

$= 60/216 + 60/216 + 20/216 + 15/216 + 1/216 = 156/216 = 0.722222$ (72%)

Therefore, the overall improvement chance is approx. 72%. He can also get nothing and retain one pair $P = {}^{60}/_{216}$. If the player attempts to lengthen the partial straight 'T,J,Q,K' by re-rolling 1d6, he can do this by getting a '9' or an 'ace',

P(convert 4 straight to 5 straight on 1d6) $= {}^{1}/_{6} + {}^{1}/_{6} = {}^{1}/_{3}$ (33%)

He does of course have a chance of getting a higher pair $P = {}^{4}/_{6}$ so we have an example of $P = 100\%$ overall, $({}^{2}/_{6} + {}^{4}/_{6})$, but actual improvement is ${}^{5}/_{6}$.

That is, if he fails to get the '9' or 'A', he would need to get a 'J, Q or K' to improve on his original pair of '10s'. A straight, especially if a **royal straight**, is ranked higher than 3oak (as with casino poker dice), but overall keeping the pair has more potential. The latter example has one re-roll and calculations that are more complex would be required to encompass a second attempt.

Players can also modify their choice of re-rolls depending on the number of opponents and their position in the series of turns. If you are at the end of row, you can see what all the others have attained, e.g. the highest hand so far is 3oak - you need a straight or more. The other way to look at it is to consider the expected value on the first roll according to number of players, e.g. with 10 players some expected values are: 4oak = <1, pair = 5, two pair = 2, treble 2 ... If you have two pair on your first roll and there are no higher rolls to beat then, roll the odd die to try for full house (33%). If there is a full house or straight out, then you could keep the higher pair and risk 3 dice to get 4oak or 5oak. With only 2 -3 players, even a pair is expected about once on average and all the others <1, so two pair starts to look like a reasonable hand.

Yahtzee®

The dice game *Yahtzee*® and related games (*General, Yacht*) are similar to poker dice in that they are played with 5d6 and have similar dice configurations and some of the same initial probabilities. Points are scored (based on pip count in the 'upper section' of the score card/table, some fixed in the lower) for the various patterns, with up to two re-rolls allowed. This game's objective is to score more points than opponents' but each scoring pattern can be used only once, e.g. once you score with a full house that slot cannot be used again in the game. Therefore, you need to target other patterns, although a higher pattern can be used to fill a lower compatible slot (otherwise 0 points), with appropriate adjustment of the score.

Calculations are more complex for some **initial roll** outcomes, with some categories merged and a special '4 out of 5' straight. Allowed patterns depend on the game variation. Thus, basic pair, two pair, full house, 4oak and 5oak (a 'yahtzee') and a straight have the same probabilities as illuminated above for poker dice (Table 20.1).

There are some variations from the standard probabilities. Like poker dice, the more probable outcomes are pairs and two pairs, etc., producing a common event:

$$P('pair\ OR\ two\ pair') = 0.462963 + 0.231481 = 0.694444\ (69\%)$$

This event union with one pair can also contain a 'small straight' (see below). Unlike the poker style game, any roll can score in *Yahtzee®* provided there is a suitable slot. Thus, the above, ideally of higher scoring pair(s), can provide a stepping-stone to trebles and quads, as they score on <u>all</u> dice. Low scoring pairs may be sacrificed in favour of a single high pip die.

Any pattern that is 'spare' can be allocated to a 'chance' category (sum of the pips). A full house can be scored alone or can be classed as a treble, hence,

$$P('treble\ inc.\ full\ house') = 0.154321 + 0.0385802 = 0.192901\ (19\%)$$

A 'four out of five' straight (referred to as a *small straight* in *Yahtzee®*) is possible as a scoring pattern in some game versions and they can also be a stepping-stone for attainment of a full straight on a re-roll of one die. A small straight with no pairs = 1 1 1 1 0 1 would give $6!/5! = 6$ forms but we only want the ones where 4 are in sequence = (1111) with one other value out of sequence. There are but two cases 1,2,3,4 with '6 and '2,3,4,5,6 with '1' in 5! = 240 ('2,3,4,5' will have a '1' or a '6' to give a large straight) which results in,

$$P = 240/6^5 = 0.0308643\ (3\%)$$

The small straight with a pair will have a repeat of one member, thus there will be four possible forms for each of the above giving 12 in total, in 5! /2! = 60 = 720 total plus the non-pair versions = 960 and,

P('*small straight*') = 960/6^5 = 0.123457 (12%)

If the straight is excluded in the game then it is included in the 'nothing' roll (480 + 240 = 720/ 6^5) and,

P('*nothing inc. straights*') = 0.0925924 (9%)

Although the latter pattern does not score specific points, it, and some other forms, are useful for calculation of re-roll probabilities in such games. Some of these probabilities are summarised in Table 20.1.

Strategy in *Yahtzee®* is complex due to the many branches in the choices (www.yahtzee.org.uk) but we can give some guidance. Optimum actions relate to maximising the expected score during choice at every stage. Generally, ensuring as full use as possible of high pip counts in the upper section of the score table, rather than over concentrating on the fixed-point patterns. These former slots score more for higher face values, so try to get them, e.g. five '6s' score 30. For example, if you get '6,6,6,2,2' early in the game, your expected score is higher if you use the treble for the '6s' rather than grab the full house. This is important, as you get a bonus once you reach a break point score (63). Balanced against this, low scoring and low probability slots can be sacrificed to ditch unwanted patterns near the end of the game, e.g. the '1s' slot score a maximum of 5 - a yahtzee (5oak) scores 50, but its probability is the lowest (5% on 3 rolls). Some patterns have higher probability and can be left until later in the game, such as full house (≈59%) and small straight (≈50%). Elucidating these probabilities requires very lengthy (or complex) calculation of the many pathways or use of simulation. Targeting a pattern can involve tactical changes, e.g., when trying for full house from 3oak, keeping the treble plus one odd die improves the conversion probability (14% cf. 17%, although this excludes the possibility of 5oak). Keeping a record of the frequency of a win by these modes of play, or the final score in solitaire play, would allow a simple assessment of effectiveness (more comprehensive simulations appear in the literature (Verhoeff; en.wikipedia.org)).

Farkle

A similar game to the above, in respect of face value patterns, is *Farkle (Ten Thousand),* which uses 6d6. The methods above are extended for the extra die, but there are complications, making *Farkle* an ideal example for illustrating some unusual probability calculations. The game is different in another way, as it is a type of 'jeopardy' game where a decision needs to be taken to retain an intermediate score or risk further throws in the hope of making a higher number of points. In addition, initially thrown patterns cannot be improved once retained. Scoring dice are selected and the remaining ones thrown, so it is a diminishing pool of dice with consequential effects on probability. In these games, '1s' and '5s' score points in any number of occurrences; this means there will be at least one pair with 6d6 if '1' or '5' is excluded, e.g. '1,2,3,4,6,x' – the last value must be a repeat of one of the previous if '5s' are excluded, so any pair will score. Exclude both '1' and '5' and the minimum scoring pattern is a treble, as 'two pair' on the first throw (except along with '1s' and '5s') does not score, being a 'farkle', which terminates your turn, e.g. '2,2,4,4,6,3' (non-scoring patterns on later throws also result in a farkle). However, it is theoretically possible to continue a turn until you reach the winning total (10,000 or more), provided you continue to score with all remaining dice.

Table 20.1 gives the counts for **initial roll** probabilities for some of the patterns with 6d6, so '6,5,4, and 3 of a kind', 3 pair and 2 trebles have these general probabilities, but the exclusion of '1s' and '5s' in some cases complicates matters. In the Farkle game, straights must be full, thus only one form is possible, i.e. 6 consecutive numbers and probability is as for 5d6 + 1 die,

$$P('6d6, \textit{full straight})' = \frac{1}{6}^6 * 6! = 0.0154321 \ (1.5\%)$$

Generally, for the first throw with all the dice, 37% result in a treble or more and other higher scoring states push this up to 42%. When sets with '1s and 5s' are added in, overall success chance with 6d6 is high (98%). Formula (18.9c) on named and unnamed mixes provides the solution route to probabilities for these latter patterns.

Thus, a roll of a single '1' with other any non-scoring values, (so '5' is excluded), must have at least one pair, as all unnamed cannot be distinct, e.g. '$1,v,w,x,y,z$' (one pair from v to z are equal),

P('6d6,1 + *any four distinct (except 5), one paired*')

specific case $\longrightarrow$ '1,2,3,4,6,3' gives,

$$= \frac{1}{6}^6 * {}^4C_1 * {}^3C_3 * 6!/2! = 4 * 360 = 1440/6^6$$

The catch here is that you cannot use face value '5'. There are four ways to choose the value to be paired then all 3 remaining are required for singles. The first section gives the unordered state and the factorial part produces the arrangements; the multipliers are 'choose the 4 unnamed values' and then 'choose 1 to be paired' $= {}^4C_4 * {}^4C_1$.

The other combinations of a '1' with another five non-scoring dice come as the two pair version of the above,

P('6d6,1 + *any three distinct (except 5), two paired*')

specific case $\longrightarrow$ '$1,2,3,6,6,3$',

$$P = \frac{1}{6}^6 * {}^4C_3 * {}^3C_2 * 6!/2!2! = 4* 3 * 180 = 2160 /6^6$$

This time only 3 out of the 4 are selected then 2 of these are taken for the pairs. The two probabilities are added via the addition rule ('OR') to give:

P('6d6,1 + *any three or four distinct (except 5), one or two paired*')

$$= 3600/46656 = 0.0771605 \ (8\%)$$

The same calculations apply for '*a 5 + any, excluding 1* ' and we can get '*a 1 or 5 with any other non-scoring*' by doubling the above,

$$P(\text{'6d6,1 OR 5 + any three or four distinct (except 5 OR 1),}$$
$$\text{one or two paired')}$$

$$= 0.154321 \ (15\%)$$

The 'distinct' condition for the unnamed means that no other scoring dice such as trebles or quads can occur and for any unnamed except '1', but these can be calculated in a similar manner (see Problems & Exercises).

Getting a single '1' and '5' together in the same throw, as with '1,5,w,x,y,z' would give a straight if the unnamed 'w to z' differed from each other and the named, so for a non-straight state, there must be a pair or two pair, etc. as above,

specific case $\rightarrow$ '1,5,6,4,2,6',

$\quad P(\text{'6d6,1 AND 5 +any 3 distinct, one paired')}$

$\quad = \frac{1}{6}^6 * {}^4C_3 * {}^3C_1 * 6!/2! = 4 * 3 * 360 = 4320 /6^6 = 0.0925926 \ (9\%)$

$\quad P(\text{'6d6,1 AND 5 +any 2, both paired')}$

$\quad = \frac{1}{6}^6 * {}^4C_2 * {}^2C_2 * 6!/2!2! = 6 * 1 * 180 = 1080/6^6 = 0.0231481 \ (2\%)$

Probabilities for occurrence of '1,5' once, twice and three times along with <u>any</u> other elements except '1 and 5', are 16%, 3% and 0.04%, respectively.

Getting a single '5' on first throw has lowest score (50 points; $P = 8\%$) and the highest is 'six 1s' (5000; $P = \frac{1}{6}^6 * 6!/6! = 0.0000214335$). A 'farkle' on 6d6 occurs as 'two pair', excluding use of '1s and 5s', and is obtained by modifying the calculation above for 'two pair',

$\quad \rightarrow$ '3,2,6,6,4,4',

$P = \frac{1}{6}^6 * {}^4C_4 * {}^4C_2 * 6!/2! = 1 * 6 * 180 * 12 = 1080/6^6 = 0.0231481 \ (2\%)$

Later rolls in Farkle will involve fewer dice and probabilities are adjusted accordingly and follow those for *Poker dice* with trebles (inclusive of pairs) and mixtures of '1s' and '5s' being more prominent (Table 20.1; see example in Problems & Exercises and Section 18.4).

The probability of getting a **farkle** depends on the number of dice thrown. As mentioned above, at the start (6 dice) it is 2%, then it increases, from 5 dice to 3 dice as 8%, 16%, and 28% then 44% and 67% for 2 dice and 1 die, respectively. The lowest score possible is one '5' over three throws. Getting this on 2nd and 3rd throws (to give the lowest possible total when reaching 3d6 (farkle chance is now 28%) is equivalent to 'a single 5' on 6d6, on 5d6 and 4d6, all with non-scoring elements. These were calculated (not shown) and,

$$P = 0.0771605 * 0.13 * 0.185 = 0.00187433 \ (0.2\%)$$

Part of the **strategy** in *Farkle* relates to the jeopardy feature – when do you stop in a turn and take what you have? It becomes a matter of weighing up the expected value for gambling with the points you have, against the possible gain on continued rolling.

The decisions begin when you select at least one scoring die from the first throw (6d6). The risks here at 5d6 and 4d6 are low (farkle probabilities 8% and 16%, respectively) and for any score you get with 1 or 2 dice, it is safe to continue (i.e. there is a positive expectation). The break point comes with 3d6 (farkle 28%), where you can have much higher scores on the first 3d6 and possibilities of even more on the second 3d6. A fuller work out of the expectation can be used to calculate the maximum score that you would risk (Z) as,

$$E(X) = - Z * 28/100$$
$$+ 50 * 48/216 + 100 * 48/216 + 100 * 12/216 + 200 * 12/216$$
$$+ 150 * 24/216 + 200 * 3/216 + 250 * 3/216 + 1000 * 1/216$$
$$+ 200 * 1/216 + 300 * 1/216 + 400 * 1/216 + 500 * 1/216$$
$$+ 600 * 1/216 + 452 * 12/216$$

$$= - Z * 28/100 + 111.92 \approx 112$$

The main part of this expression calculates the expected score on 3d6 ($\approx$

87) and the last part covers the possible re-throw of all six dice with the maximum scores for 4oak or more and special patterns (the average for this depends on the scoring variations used and can be worked out as for 3d6 or a simulation used to count them). For the level of risk that results in a fair game, the magnitude of $(Z * 0.28)$ must equal 112 and for this, $Z = 112/.28 = 399.7 \approx 400$. Risking any more points would produce a negative expectation. Similar calculations give the limit for various nos. of dice and for 2d6 and 1d6, maximum risk levels become markedly lower (< 300) and only in the desperate state of an opponent nearing the line would such a risk be worthwhile. If you do score with all six, you can throw all six again but now the risk level changes as you will have more points in the current turn; the limit here is safe for 6d6, with the danger increasing for fewer dice.

In addition to these considerations, you can try to improve your initial score if it consists of lower scoring singles such as '5s' or even a low treble ('222') depending on how far ahead opponents are. For example, you roll 'two 5s' on 6d6– keep one '5' and reroll 5d6; more risky would be to keep a 5' or '1' single but abandon any accompanying low scoring treble ('three 2s') in an attempt to get more with 5d6. More sophisticated treatments than the above can take into account effect of 'triple farkle' penalties and extra rolling where possible and various scoring levels (www.math.cmu.edu; www.bluffton.edu).

20.2.2 Counts and magnitude of face values

In the above games, configurations that showed patterns were important but the number of spots on the dice did not always matter, e.g. a yahtzee on 5d6 can be achieved with any face value. Other games still note the number of dice but also compare magnitudes.

Risk®

In *Risk®*, dice are used to decide battles between battalions using 1 to 3d6 (the attacker), against 1 to 2d6 (the defender). Face values that are equal or higher decide the issue, just like the situation described for competing numbers of dice (Section 15.2.1). Attacking and defending probabilities and the number of battalions lost, etc. can be calculated but we will concentrate on the attacker win probability, where the attacker must beat each defender die.

To get these we could look at all sequences (up to 7776 of them) for 2d6 to 5d6 and total the cases where the attacker wins, etc. - a search exercise for a computer program, which indeed is one way to determine these probabilities for the game. The simplest battle is with 1d6 for each opponent and the probability of a win for the attacker is illustrated in Section 15.2.1 and P = $^{15}/_{36}$ (41.6%; a tie signifies a successful defence). The probabilities for battles with more dice are more difficult to attain and require use of the *multiplication* and *addition rules* ('AND' and 'OR', respectively) along with counting techniques, some listing and use of the *binomial function*. In this game battle system, there are no ties or draws, as these are counted as a failure for the attack.

If we take a simpler battle, where an attacker uses 2d6 and the defender uses 1d6, a fuller account illustrates the principle. If the defender rolls a '1', then to beat this and win, the attacker needs roll at least one instance of '2 or more' on the 2d6,

$$1-BINOM.DIST(1-1,2,5/6,1) * 36 = 35$$

(Alternatively, using the complement we can see that only one outcome out of the 36 on 2d6 result in a draw ('1,1') so the remaining 35 win).

We can calculate the count in this manner for all individual cases, but a neater way is to sum the binomial functions by applying the OR rule for the defenders rolls '1 to 5' (when defender rolls a '6' the attacker cannot win),

P('*attacker win, 2d6 vs. 1d6'*)

$$= SUM[(1- BINOM.DIST(1-1,2,d/6,1) * 36)]/216$$

for d (defenders roll) = 1 to 5

$$= 125/216 \ (57.87\%)$$

Unfortunately, this can't go in Excel and each binomial function must be listed then summed as per spreadsheet output:

Spreadsheet Table 20.1

Att.	Def.	Def. gets	AND	Att. needs on one of 2d6	P	Formula
2	1	1	AND	at lst. '2'	35/216	1/6 * (1 -BINOM.DIST(1-1,2,5/6,1))
		2	AND	at lst. '3'	32/216	" ,4/6, "
		3	AND	at lst. '4'	27/216	" ,3/6, "
		4	AND	at lst. '5'	20/216	" ,2/6, "
		5	AND	at lst. '6'	11/216	" ,1/6, "
					125/216	0.578704 (57.87%)

The same procedure can be used for 3d6 vs. 1d6 (65.97%). The case for 1d6 vs. 2d6 is simpler. The defender's 2d6 outcomes are listed, and the corresponding attacker 1d6 outcomes that can beat each are stated and the probabilities summed. For example, if the defender rolls up to a '2' on either die, the attacker must roll '3 or more' to win on the single d6. These enumerate to $^{55}/_{216} = 0.254630$ (25.46%).

When the competing dice are at 2d6 each, the binomial function can be used again, although not as simply as above, as it over counts some outcomes that do not qualify. One way round this is to use the function then get the over-count by simple manual counting. The various values can be put into the function as was done for 2d6 vs. 1d6, but in this case, all can be reduced to a single equation,

P('*attacker win, 2d6 vs. 2d6*')

$$= \text{SUM}[((6-d1)*(6-d2) +(d1-d2)*(6-d1)) * (2 - (d1=d2))]$$

For d1,d2 = 1,1 to 5,5

Where, d1, d2 are the defenders roll entered as equal or as the higher

These are summed to give the probability = 0.227623 (22.76%)

For 3d6 vs. 2d6, the possible sequences need to be broken down more for same cases with dual functions used. The pairs on the defender's 2d6 are simplest and are given by,

$$P = (1 - \text{BINOM.DIST}(2-1,3,(6 - d1)/6,1)) * 216$$

Where, d1 = the first (or second) of the defender dice value

This counts 'at least two of the face values above d1. For example, to beat '4,4' on the defender's roll, d1 = 4 and the calculation is,

$$= (1 - \text{BINOM.DIST}(2-1,3,(6-4)/6,1)) * 216 = 56 \text{ cases qualify.}$$

Others are a bit more arduous with often a 'blow by blow' process, examining for criteria. Take the calculation for the attacker to beat '4,1'.

$$P = (\text{BINOM.DIST}(2,3,4/6,0)) * 216 - 6$$
$$+ (1-\text{BINOM.DIST}(2-1,3,2/6,1)) * 216 = 146$$

The first binomial expression counts the sequences with exactly two occurrences of '1-4' combined with a '5 'or a '6,' e.g. as with '3,2,6' or '1,4,5' – this enumerates to 96 but 6 sequences (those with '1,1,x') cannot beat '4,1' and are subtracted. The second expression counts each sequences of at least two of '5-6, with '1-4, all of which beat the defenders dice. The other counts proceed as above but the counting of non-qualifying cases gets more difficult for some. These are summed to give 2890/7776 = 0.371656 (37.17%).

Similar calculation procedures can be used to elucidate the probabilities for winning and losing part of the exchange such as when the defender, attacker, or both lose a battalion, etc. Of course, a single exchange as above may not be the end of the battle and further attrition may occur. Some of these can be strung together to give assessment of the end win probability, such as that for the attacker gaining the territory. These are just some of the ways in which the engagement can be fought and there are more encompassing measures on the web (www.datagenetics.com). *Dice Wars* games procedures are similar to *Risk®* but the dice sum is used rather than the face value comparison (see below).

Pass the pigs

One final game with a dice structure and use of 'face' values is ***Pass the Pigs***©David Moffat Enterprises. Related to the game of ***Pig*** below, the play pieces are miniatures of the pig animal. There are two of them and when thrown or tossed the toy pigs land in one of a number of positions, on their back, on their side, etc. These represent the 'face values' There are no 'true' probabilities for these outcomes and they have been determined by relative frequency experiments from a number of sources (values range from about 35% for a side landing to approx. 1% for the jowls (a' leaning jowler'), with trial numbers ranging from several hundred to several thousand (Kern; passpigs.tripod.com). This is perhaps the ultimate example of non-equiprobability for the numerous outcomes, some of which are very usual, such as the pigs landing on top of one another, etc. Scores are updated with a running total and the first to a nominated level (100 or 500) wins. During a turn, some events cancel the turn or cause a 'lose-game' state. Depending on the version, a player can make a decision to continue and a turn continues until stopped, so a consideration of payoff is useful. Apart from this strategy, the game is a bit more complex than *Pig*, although it appears to be more of a fun game, watching the pigs getting into the various configurations, etc.

Combat and other actions with dice face values

With war and role-play / adventure gaming where counters, tokens or miniatures are used to represent the units, combat and some other actions are decided by face value magnitude and occurrence. Actions are allocated a particular face value on a single die to achieve or exceed in magnitude. Getting the probabilities is straightforward except in cases where an inequality is used. Sometimes, when the play involves 'non-d6 dice', the probability is not quite so obvious and the methods of Section 18.2.1 are useful. Some examples in various role-play, adventure and board games are:

'You need at least 11 on 1d14 to lasso that branch', $P = 14-11+1 = {}^4/_{14}$

'The floor is now slippery there is a 1 in 4 chance that you will slip- roll '2-4' on 1d4 to avoid', $P = {}^3/_4$

'That's an almost impossible task - you need to get at least 90 on 1d100 to succeed', $P = 100-90+1 = {}^{11}/_{100}$ (11%)

'A roll of 5 or more' (1d6) will force a retreat', $P = {}^2/_6$

'A 12 or more' on 1d20 achieves a hit' $P = (20-12+1)/20 = {}^9/_{20}$

'Roll a d8 to assess damage', $E(X) = 4.5$

'The plasma gun scores 1d10 of destruction', $E(X) = 5.5$

'73 or less' on 1d100 gives a parry, $P = {}^{73}/_{100}$

'The sling shot requires '16 or more' on d20' $P = (20-16+1) = {}^5/_{20}$

'A 4 or more on d6 causes armour absorption', $P = 3/6$

'A '9' or more on the d20 means you resist the attempt to control you', $P = (20-9+1) = {}^{12}/_{20}$

(This latter example is a 'luck' or 'fate' 'saving throw' - a very important last chance situation in some games where the player gets the opportunity to avert disaster). Other possibilities are where minimums and maximums of the dice can be designated as penalties ('fumbles') or bonuses (critical hits or 'roll another die'), respectively.

Some of these examples are true duels as described in Section 15.2.3 and we can apply the formulae that were established there. Assume the sling shot adversary faces an opponent with a cross bow ($P = {}^7/_{20}$). If the slinger has the advantage (fires first), what is his chance of winning on alternate shots (assume 1 hit stops the exchange). Using formula (15.1a) with a = $^5/_{20}$, b = $^7/_{20}$,

P('*slinger fires 1st, slinger hits*') = $^5/_{20} / (^{12}/_{20} - ^{35}/_{400}) = 0.487805$ (49%)

The opponent archer appears more adept or experience and his chance (formula 15.1c),

P('*archer 2nd, archer win*') = $^7/_{20} * (1 - ^5/_{20}) / (^{12}/_{20} - ^{35}/_{400}) = 0.512195$ (51%)

The crossbow archer fires second, but because of better ability has a slight edge, but the exchange is close, and the opponents could consider other options. The other melee actions above could be analysed in the same way if opponents are involved either with ordered strikes or simultaneously.

Another feature of the above combat actions is that they all employ single dice but others systems are based on multiple dice, as mentioned above and discussed below. Counting of the incidence of face values is found in some systems. One game type (www.white-wolf.com), uses rolls of multiple d10s where any face over a target value is a hit or a success, but '1' can cancel hits etc. Some have re-rolls on '10', i.e. another die can be rolled. Nos. of dice vary, so if there is an action where '7 or more' is required the obvious solution would appear to be use of the binomial function, e.g. '*at least one 7 or more on 5d10*',

P = 1-BINOM.DIST(1-1,5,4/10,1) = 0.922240 (92%)

However, this includes the presence of '1s', so this is not the correct probability for success. Taking this into account is difficult by direct calculation, but a simulation is relatively straight forward to set up. RANDBETWEEN(1,10) is entered in 5 columns then the incidence of '1s' and successes are counted (COUNTIF( )), then totalled. Successes are counted by a simple comparison of the greater for a win = IF(no. 1s > no success, "", "WIN")). This gives an estimate (1000 repetitions) of 0.76 (76%). This can be expanded for any possible combination of nos. of dice, difficulty levels etc., plus further analysis. Advantages are that with several dice then there can be 'degrees of success' or failure, allowing a range of possible effects to be implemented in running of the game. Different sizes of dice can be used to give variation in difficulty, as well as a more difficult target value, e.g. the same task on d8 vs. d10 vs. d12 ...

Other war games deal with larger groups of combatants from war-band, platoon up to regiment, etc., depending on the setting. Dice are used in variety of ways to decide combat results. For face values, a common system uses one or more dice (usually d6) as one per combat unit, and then each opponent rolls the number appropriate to their force. Hits (successes) are awarded by getting a certain face value or more. For basic units this could be a 6 on 1d6, elite units may need only a 4 or more, i.e. the initial roll can be modified by skill and experience, and other factors such as weapon strength, positional advantage, etc. Dice achieving the required hit value are separated and counted. Further rolls, one for each success, are modified by other effects, such as defender's cover, armour, etc. and decide damage and disrupt, retreat, rout or elimination effects.

The +/- effects influence the probabilities, e.g. an elite unit may inflict hits with high probability but if the defender has strong armour and is in a good defensive position, then the probability of damage may be relatively low. Compare achieving a hit with 4-6 (50%), with the probability of damage on a weak unit in open, requiring 3-6 (67%), a good chance, with a defender who has strong armour and is in cover causing -3 on the aggressor's roll, requiring a 6 (17%), very poor odds, to score damage.

Systems that employ such methods are rapid and large numbers of units can be involved in a battle with the binomial probability function coming to the fore for calculation:

Fifty units (force A) attack a force of 30 units (B). A rolls 50d6 and B rolls 30d6. A's units are partially trained and require a '5-6' to score a hit. B's are super elite troops and can get a hit with '3-6'.

Is it a sure win for A, or can B beat the superior (in numbers) force?

What is A's chance of winning in one exchange?

Assume simultaneous attacks and that a 50% loss on the first exchange will cause a rout, without considering damage. If A scores 15 or more (successes), B suffers a 50% elimination:

A's attack to get 15 or more hits,

$$P = 1\text{-BINOM.DIST}(15\text{-}1,50,2/6,1) = 0.738761 \ (74\%)$$
$$(10 \text{ or more}, P = 0.987292)$$

B's attack to get 25 or more hits,

$$P = 1\text{-BINOM.DIST}(25\text{-}1,30,4/6,1) = 0.0354542 \ (3.5\%)$$
$$(15 \text{ or more}, P = 0.981205)$$

The situation is just too extreme for B to win on one exchange but A has a good chance. The probabilities increase sharply for lower numbers of hits and assuming both survive, A stands a good chance of reducing B to 20 or less units with B almost certainly wiping out 15 or more of A's force. The next exchange would be at A 45 vs. B 20 (B's chances are abysmally low but A would have a 96% chance of getting 50% elimination). Maximum exact probabilities are at the mode for each opponent's units, namely 20 successes for B at 15% and 16-17 for A at 12%.

20.3 Dice sum probability and game applications

In this section, we consider dice use in games where the magnitude of a single die or the *sum* of two or more is the deciding factor (some single die cases were dealt with above). D6s predominate, but any size can be used with standard or non-standard numbering. Totals are calculated for the dice but additional 'bonus' dice and special outcomes boost the original roll. Designations like 'exploding dice' and 'speed dice' (used to augment play by extra rolls or extra dice, respectively) have given rise to a wide variety of format and complexity.

Pig

The game of **Pig** (*Piglet* or *Jeopardy*) is based on 1d6 in its basic version. Each player rolls 1d6 one or more times and keeps a running total of the sum attained each turn. The player keeps going until a voluntary stop is made, or a '1' is rolled, which loses all points for that turn. First to 100 wins and the main question on play strategy relates to 'when to stop?' To answer, we can first work out the probability for survival at any point. These are calculated by,

$$P(\textit{'survive on roll no.'}) = \text{success probability}^{\wedge \text{ roll no.}}$$

Overall, the probabilities start at 83% on the first roll ($= (^5/_6)^1$) and reduce correspondingly as more rolls occur. For example, to survive up to the 7th roll has probability of $(^5/_6)^7 = 28\%$. These can be confirmed by the binomial function, e.g. 'survival by roll 3' is equivalent to '*getting a 2 or more on each of 3d6*',

$$P = \text{BINOM.DIST}(3,3,5/6,0) = 58\%$$

As a rough guide to strategy, stopping at the fourth roll, where the probability is just under 50%, is one guideline. Being more conservative would hold at 3 rolls, and a bit more risk could run to 5 or 6 rolls. Each roll, if it succeeds, gains '2,3,4,5 or 6', average $= 20/5 = 4$, as you can only gain on points 2- 5. Thus, the '4 rolls' rule above will, on average, produce:

$$E(X) = 20/5 * 4 = 16 \text{ points} \pm 3$$

One or 2 rolls more gives around 20, which agrees with the figures found by some more complex mathematical analyses (Neller and Presser). Another way to look at the play is the payoff choice of 'bet' against what you have of getting more. As your current total increases, there is less payoff and eventually the return becomes negative. For example, you have 12 points - do you continue? If you go on, you are in effect betting your 12 points in the hopes of winning more. The expected return is,

$$ER = - 12 * {}^1/_6 + 4 * {}^5/_6 = -2 + {}^{20}/_6 = -2 + 3.3 = + 1.3$$

On average, you should still gain. This return reduces and at just over 20 points, it becomes negative,

$$ER = -21 * {}^1/_6 + 4 * {}^5/_6 = -{}^{21}/_6 + {}^{20}/_6 = -3.5 + 3.3 = -0.2$$

Adherence to these stopping points also depends on where you and the opponent are on the scale up to the 100 mark. If your opponent is well ahead, especially near to the win level then your only chance may be to keep rolling to catch up and hopefully overtake. The factor to take into account is the score attained and use that as a marker for the stopping point. Making up points by rolling more than this and surviving longer, gets increasing remote in terms of probability. The minimum no. of rolls is 17, i.e. seventeen '6s' take you to 102 points. Unfortunately, this is not going to happen much(!) as,

$$P(\textit{'seventeen 6s'}) = 0.0000000000001$$

Assuming average sums and similar tactics, whichever player starts first has an advantage, but *Pig* is a very dynamic game with circumstances changing quickly.

Craps

Craps is played with 2d6 and in terms of probability, the game illustrates at least two aspects due to the nature of the scoring system based on the sum. Initial probabilities are provided by the sums for dice sequences of 2d6 (Table 4.5, Book One) and others as shown in Table 20.2).

Table 20.2 Craps: Initial roll probabilities for sums

Sum	No. sequences	Probability	Game result
7 or 11 ('natural')	6 + 2 = 8	8/36 (0.222222, 22%)	a win, 'pass
2,3,or 12* ('craps')	1+2+1 = 4	1/9 (0.111111, 11%)	lose
4-6 or 8-10 ('point')	3+4+5+5+4+ 3 = 24	2/3 (0.666667, 67%)	continue, 'point'

*in casino craps '12' = 'push' ('no win')

A roll of '7 or 11' wins outright, '2, 3 or 12' lose and any other sum total is called a *'point'*. This latter result opens the door to continued play and the player must attempt to convert this to a winning result. The player throws again in an attempt to get the point sum again. There is no restriction on the number of throws to get this but a '7' terminates the process. A win can now be achieved in several ways and the probabilities are given by some counting then the summing of a geometric series.

For example, a point roll of '9' can be converted to a win by rolling a '9' (4 ways), but a '7' (6 ways) loses immediately and any other outcome (26 ways) allows another attempt. Similar reckoning can be done for the other point values revealing that '6' and '8' have the highest probability of being converted on the first re-throw (5 ways) and 4 and 10 are least likely (3 ways). If the player fails on the first attempt, they can keep rolling to infinity if wished until stopped by a fail or a win. In practice, much fewer attempts are required but can still reach high incidence. The average number is obtained by the geometric E(X) (1/p), Section 19.3.3. For a point of '9', the conversion probability is calculated as described below, to give $p = (^4/_{36}$ * $^4/_{36}) / (1 - {}^{26}/_{36}) = 0.044444$, hence,

$$E(X) \text{ 'no. of rolls to convert point ' 9'} = 1/0.044444 = 22.5$$
$$\text{(point of '10' has 36)}$$

More are quite possible and 154 rolls are quoted on the web in one casino (www.nextshooter.com). The overall probability for the conversion is calculated by applying the method detailed in Book One Section 10.2 and in Section 15.2.3 (Spreadsheet Table 15.3). As seen, this is an 'ever decreasing' series, in fact, a geometric series as described in these latter sections. Using the technique given there, e.g. for a particular point roll, say '4', we can calculate the win roll probability.

The first term is $^3/_{36}$ * $^3/_{36}$ and the common ratio in the series is $^{27}/_{36}$.

Hence, using (10.1c),

$$\text{Sum} = {}^3/_{36} * {}^3/_{36} /(1 - {}^{27}/_{36}) = {}^9/_{36}{}^2 * {}^{36}/_9 = {}^1/_{36}$$

Thus, overall win = 'get 4 then convert 4 '= $^1/_{36}$ = 0.027778 :

Spreadsheet Table 20.2

Player can win by	P
P(4) * P(4)	3/36 * 3/36 = 1/144
OR P(4) * P(2,3,5,6,8-12) * P(4)	3/36 * 27/36 * 3/36 = 1/192
OR P(4) * P(2,3,5,6,8-12) * P(2,3,5,6,8-12) * P(4)	3/36 * 27/36 * 27/36 * 3/36 = 1/256
OR P(4) * P(2,3,5,6,8-12) ...	3/36 * 27/36 * 27/36* 27/36 * 3/36 = 1/341
... to ∞	

Similar procures are required for the other point outcomes, showing that '6' and '8' have the highest probability (0.0631313) and '4' and '10' the lowest (0.027778).

Sum these and 0.27 (27%) is the chance of winning via the point, and combining this with the win by '7 or 11' gives 0.22 + 0.27 = $^{244}/_{495}$ = 0.49 (49%). A bet on this ('a pass') has expected value (return) as,

$$E(X) = -1 * {}^{251}/_{495} + 1 * {}^{244}/_{495} = -0.01414$$

However, *Craps* has a number of bet types and in one, players can wager against the above, i.e. that the shooter fails and this has a 51% chance with a positive expectation,

$$E(X) \text{ 'don't pass'} = -1 * {}^{244}/_{495} + 1 * {}^{251}/_{495} = +0.0141414$$

These calculations apply to 'street craps' or to private games as when playing at home with friends. In casinos craps, the odds are adjusted to be closer, as '12' is not counted in the fail chance and this gives,

$$E(X) = -1 * {}^{244}/_{495} + 1 * ({}^{251}/_{495} - {}^1/_{36}) = -0.0136364$$

This represents the casinos' cut per 1 unit. Similar treatments can be used to calculate the odds for the other outcomes and the many betting options in the game.

[A simpler method for calculation of point conversion probabilities is to view the reduced sample space of the conversion outcomes; with the point as '4' there are then 3 outcomes to succeed and 6 to fail (sum = '7') so the conversion chance is $^3/_9$ and overall = $^3/_{36} * ^3/_9 = ^1/_{36}$ (Packel)].

Board games

Many board games use the sum of 2d6 for a variety of actions including encounter resolution and combat (see below) but perhaps movement is the commonest. This time, the distribution of probabilities for the possible sums is not uniform and some movement distances will be more likely than others will. Sometimes, the sum value may be split amongst several game pieces for movement or other actions. Additional points can be gained by boosting the initial roll when a designated event occurs. For single dice where this rule applies, the die is rolled and if equal to the specified one such as the maximum, another roll is permitted and so on (the 'exploding dice'), until the series terminates on a failure. Higher sums can be obtained. For example with d6, the trigger is a '6' and a fail has P = 5/6. The average number of rolls for the 'exploding dice' is given by the geometric expected value, $E(X) = 1/p$ = 1/ 5/6 = 1.2 and the average sum = 6 * 1.2 = 7.2.

Instead of a roll to get the maximum sum, target sums are defined for different circumstances of the play. Doubles ($^6/_{36}$) and '7, 11' ($^8/_{36}$) are popular, one at a time or together ($^{13}/_{36}$) and rolling less than a certain outcome to limit the probability of getting out of less desirable places on the board.

Other enhancements include addition of a 'speed die' - an extra dice to boost certain rolls in the game and shorten play or action times. The process can involve allowing the player to choose the better die (the best of two if used with 2d6), or the composite result. For the effect on probability, we can use the unnamed + named methods (Section 18.3.4). Take 1d6 with use of a speed die as an extra d6 - to get *one 6* originally was $^1/_6$, now becomes *one or more 6s on 2d6'*, i.e. '6,x' with no constraints on 'x'. Using (18.9b, 18.9c),

$$P = \frac{1}{6} * \frac{5}{6} * {}^2C_1 * 1! + \frac{1}{6} * \frac{1}{6} = \frac{10}{36} + \frac{1}{36} = \frac{11}{36},$$ almost doubling the original chance.

For 2d6 and *'getting a sum of 12, two 6s'*, originally $P = \frac{1}{36}$ ($\frac{16}{216}$), now with 3d6, *'getting two or more 6s and a sum of more than 12'* as '6,6,x' ($\equiv$ 12 on best two as two 6s),

$$P('x{\neq}6') = \frac{1}{36} * \frac{5}{6} * {}^3C_2 * 2!/2! + P('x{=}6') = \frac{1}{216} * 1 *3!/3!$$
$$= \frac{15}{216} + \frac{1}{216} = \frac{16}{216},$$ almost 3 times more.

In *Monopoly*®, the most well known of such games, the board has various housing properties that are crucial to success, but on the first turn it more probable that you will miss a property location. This is based on the most common sum with 2d6, namely '7' ($P=\frac{1}{6}$) which leads to a **CHANCE®** square. By assuming this average distance, we can get an idea of what locations are most likely on the first circuit, namely 7,14,21,28, and 35 (numbering clockwise with **GO®** – location 0), ignoring the effect of doubles and redirection cards. Note that all subsequent circuits will not necessarily start at position 0 - starting point can vary for each circuit. It's possible to calculate probability for any location for any number of turns. The simple instance could be on the first turn and the first circuit of the board, to get to a particular property, e.g. one of the blue sites at location 10 and assuming no jumps and ignoring other players –

> Roll once – get '10' on 2d6 = $\frac{3}{36}$
>
> OR roll twice- get '10' on 4d6 = $\frac{80}{1296}$
>
> OR roll thrice - get '10' on 6d6 = $\frac{126}{46656}$
>
> overall, the sum of these, P = 0.147762 (15%)

Beyond this, performing the calculations for all locations gets complicated and some very advanced techniques have been applied to this game (Collins;

www.monopolynerd.com). The results from many that have applied these methods have shown that the orange properties - around location 17 and the red at 21 are the most visited.

Thus, these properties are a desirable objective – probability of getting all colours in a group is higher and the probability of other players landing on these once you have them is higher - therefore more rent. A more attainable way, avoiding the mathematics would be a simulation or a relative frequency experiment over many games, noting data on how often each location is reached– which location most frequent, etc. If the game is played regularly, players can gather data on these aspects as they play the game but simulation by computer using the techniques explained in Book One Section 14.3 will be much faster. Locations on the board are allocated numbers (0-39, 1-40). Two dice (2d6) rolls are simulated using **RANDBETWEEN(1,6)** + **RANDBETWEEN(1,6)** and a running total is kept of the positions on the board. This is adjusted back to the 0-39 scale after each circuit. Simulated over 1000 rolls (approx. 200 circuits of the board) shows a rough uniform distribution with a probability of 2.5% for each location. Of course, the game has other influences on position - some locations give jumps to others. **GO TO JAIL®** has 1 on board and 2 cards ($^1/_{16}$ chance for **CHANCE®** and **COMMUNNITY CHEST®**). If the simulation code is set to jump to position 10 (jail) when landing on the **GO TO JAIL®** square (30) and assuming the player pays the fine immediately or uses a card, then positions 17, 16 and18 get favoured slightly more. Jail location becomes the most visited at P = 4.5%. Similar adjustments can be included for the effect of other jump cases with event cards.

Other probability-aided tactics are possible such as when first allocating houses and hotels - gauge the probability for an opponent landing on one of these properties. If you cannot allocate the buildings evenly on all, then put more on the one with the highest probability. Later in the game some juggling can be done by selling houses back then re-buying and reallocating

212

them for best effect. The disadvantage is that the sell-back is only for half price and watch for an auction. Other factors affecting play such as the best properties in respect of return are discussed in the references.

War games and the dice sum

Unlike the account of war game combat via face value counts above, where large numbers of dice can be common, using the magnitude of a single die or the sum of fewer appears in many other games. In the case of *Dice Wars* (1 or 2d6 to 8d6), battles are resolved by comparing the sum of various numbers of dice possessed by the armies of the attacker and defender. The calculation procedure is similar to others used in that the number of outcomes constituting a win is counted, e.g. with 3d6 vs. 2d6:

Spreadsheet Table 20.3

Defender (2d6) rolls sum	Attacker (3d6) wins with sum	Net ways
2 (1 sequence)	>2 (216 sequences)	216
3 (2 sequences)	>3 (215 sequences)	430
4 (3 sequences)	>4 (212 sequences)	636
...	...	
12 (1 sequence)	>12 (56 sequences)	56
	Total event sequences =	6054
	P(3d6 sum exceeds 2d6 sum) =	0.778549

The probability of a win = $^{6054}/_{7776}$ = 0.778549 (78%).

The number of ways for an attacker win can be determined mentally in some cases, but eventually most of us will have to use combinatorics (Section 19.3.1) to get these counts. The total is used in the classical equation with the denominator as the number of sum sequences possible with all the dice, in this instance 5d6 = 36 * 216 = 7776. Other than writing a tailor–made macro (Section 18.3.1) that can tote up the counts with a series of program loops or more elegantly call formula (19.2a) (www.superdan.net), there does not seem to be a simpler method for these calculations. The process above can be performed for all possible pairs of dice combatants (from 2 to 6 or more). For 2d6 vs. 2d6 the probability is about 44% so the value above is quite a large increase in chance for a slight increase in numbers of attacking units (dice). Most equally balanced forces have around 45% but the jump in probability as 1 army is added gets less. Logically, the greater force prevails and the highest win probability is with the highest vs. lowest die count e.g. if lowest defender dice is 2, then 6 or more dice attack with the probability over 99%. In the long term, the expected return for the exchanges show more. Expected value calculation is given by,

ER = - dice lost * the probability of lose

+ the probability of win * dice won on one outcome

e.g. a force attacking with 2 dice vs. 2 would destroy 2 but loses 1 if it fails,

ER '2 vs. 2' = - 1 * 0.556 + 2 * 0.444 = -0.556 + 0.888 = + 0.332

3 dice vs. 2 would destroy 2 but lose 2,

ER '3 vs. 2' = - 2 * .222 + 2 *.779 = -0.444 + 1.558 = + 1.114

As imagined, there is a larger gain when have we have larger armies than the defender (wizardofvegas.com; highprogrammer.com). Gain ranges from a maximum at +4.37 (8d6 vs. 5d6) to the least at -3.56 (5 vs. 8). Using probability, strategy relates to the obvious actions of avoiding low probability and low ER engagements. Otherwise, maximise chances in battles when possible, try to minimise losses and retain as many contiguous territories as possible.

Other war games base combat on much fewer dice and are restricted to the face value on one die or the sum of two. Such war games use a two-way matrix of odds - the *combat results table* (CRT) - with dice outcomes signifying hit / miss /retreat / disrupt/ eliminate, etc.

Two-way matrix tables of 'unit odds' are drawn up based on the number of competing combat units. Thus, if two evenly matched units engage in combat the unit odds for the battle are taken as 1:1, but this is not necessarily the probability of a win or defeat. Two of such units against 1 would be 2:1 and so on. In the case of units with differing abilities, the 'attack strength' of one can be compared with another to get the odds, e.g. a unit of 8 vs. a unit of 4, becomes 2:1, a unit with 3 vs. 7 becomes 1:2, etc. usually in favour of the defender.

The probabilities for the whole range of possible unit odds are assigned as the outcomes for 1d6 or 2d6, and usually there will be designed to match the unit odds. With a simple CRT based on 1d6, the 1:1 unit odds would be given a probability of 'a 4 or more', $P = 0.5$, for an effect on the defender:

Unit odds	Combat Results Table					
	1d6 outcome					
1:1	1	2	3	4	5	6
Effect	att. ret. 1	--	no effect	def. ret. 1	def. ret. 2	def. elim.

As seen, the dice range allows allocation of several possible results. By the time the unit odds reach 3:1, the attacker is usually fairly confident of at least pushing the enemy backward (e.g. '2 or more' on 1d6, $P = 83\%$. Using a single die means that the distribution of outcomes is uniform – each has equal probability. One d6 is the usual and this means that if the above effect can occur via the one roll on both antagonists that there is no convenient 'middle' value with equal numbers of outcomes on each side. This would be possible on other dice such as a d7, but as unit odds are rounded off in favour of the defender, this is balanced up by more effects on defender. Larger dice are possible giving more effect categories, such as 1d8, 1d10 and 1d12 and 1d100.

Other systems can use a combat differential – the difference between the attacker and defender's combat strength with probabilities allocated in a similar manner. Others add up combat factors for opponents, and then each rolls 1d6, adds these then the totals are compared. With these systems, the probabilities are not immediately obvious but higher die rolls boost any superiority or make up for weakness in the attacking units. Some cavalry units (+4) attack an equal number of foot soldiers (+2) – the cavalry have the advantage at +2. Each rolls a d6 - what is the probability of a cavalry (C) win against the foot (F)? This is answered by referring to the 2d6 distribution of sums except that the probability shifts up for the cavalry – with no advantage they would each have a 50% chance of win - the +2 raises the cavalry chance:

C	F		
roll (+)	**roll to lose**	**C win**	
6(8)	6 - 1	6 ways	
5(7)	6 - 1	6 ways	
4(6)	5 - 1	5 ways	F can draw with '6'
3(5)	4 - 1	4 ways	F can win with '6', draw with '5'
2(4)	3 - 1	3 ways	F can win with '5-6', draw with '4'
1(3)	2 - 1	2 ways	F can win with '4-6', draw with '3'

Thus, there are 26 ways for C (cavalry) to win, 6 ways for F (foot) to win and 4 draws out of 36 (sample space for 1d6+2 and 1d6 = 36), hence,

P(*'cavalry win'*) = 26/36 = 0.722222 (72%)

In this example, the players can take it in turn to roll, or roll together as there are multiple win states, like the example in Section 15.2.1.

With CRTs based on 2d6, the matrix will have may more segregated effects: retreat 1 to 3 / rout / advance, etc. and there can be individual sums or a range. The distribution is the triangular and probabilities are higher for sums near the centre. Consequently, if a particular effect is allocated to '7' (the mode and mean, P = 17%), then this will happen more often than others will. This can be overcome by allocating a range for other effects, e.g. 5 - 6 = 9/36 (25%) and 9-11 = 6/36 (17%), etc. With <u>each</u> player rolling 2d6 within the one combat, the number of possible outcomes increases (4d6), widening the scope of effects further.

Typically, as the attack ability of an attacker increases in relation to that of the defender, the odds of winning a battle or skirmish, etc., increase also. More eliminations will be achieved with higher probabilities and players strive to gain the best odds.

In addition to attack strength, units can be assigned values for defence, movement, speed, etc. as well as +/- for various states. Dice can be used for a variety of other actions and effects such as morale rolls, save throws, missile, artillery and siege and demolition effects, etc. depending on the era the game is set in. Six sided dice are common but percentage dice, d10, d12 and even use of a random number generator on calculator or computer or other are used in some systems.

Role-play games and sums

As seen in previous examples, dice are used in role-play games (rpg) for a number of purposes. There are many, many rpgs resulting in equally numerous dice systems and each has its devotees and detractors. Arguments revolve around suitability in mathematical terms, adherence to real activities, etc. but with some it comes down to personal preferences. One prominent point of contention relates to distributional nature (typically the uniform distribution of single dice vs. the triangular or bell-shape forms of multiple dice), but there are other issues. Overall, there are systems that use single dice and others that use various numbers of just about every size of dice available. Effects are judged on magnitudes of face values, sums and even patterns within the many systems in use. We will not get too deeply involved in this on-going debate as there are multitudes of references (Morgenson [b]), but will highlight some of the main types and features.

Determination of abilities

At the start of an rpg, the characteristics of player characters, such as physical and mental ability, etc. are often determined by rolling dice. The idea is to create characters with a range of abilities and perhaps to guide specialisms of career and activity in the game – e.g. ideally, a warrior is strong and dextrous, a space ship officer needs high technical skills, etc. This can be determined at game start by taking the sum of several dice (typically 3d6), but single die ability generation has been used. Multiple dice produce a distributional shape similar to that of actual human abilities, as in the case of human IQ scores (a normal distribution), but whether or not this is carried through in the game or if it applies to other abilities is not clear. With multiple dice, the probability of getting all abilities high is low - viz. for 5 abilities - getting all 5 above average (say > '12' on 3d6) would be $\left(^{81}/_{216}\right)^5 = 0.0074 (\approx 1\%)$, so such a character would occur once in every hundred. Achieving one exceptional ability ('18' on 3d6) would be $^1/_{216}$. If the players are rolling for 6 abilities the expected value of '18' is once in about 40 throws of the compound experiment (i.e. 40 times 3d6) – if there were 5-6 players, one '18' might appear. For these reasons, extra dice or points are sometimes included to boost the abilities for the adventurers who are considered special, raising the average value. Once determined, the abilities levels can result in modifiers for actions in the game, like how well or how poorly the character shoots a bow or decodes an encrypted star-ship battle protocol, etc. Alternatively, the ability roll itself represents the target roll and the character must roll above or below for success. Other systems allocate one or more dice per ability and during play they are used to get an on the spot value for comparison.

General actions

Like the latter instance above, all activities in the game can be gauged by dice rolls. Attempting various tasks – movement barriers, obstacles, jumping, climbing, learning magical procedures, flying a spacecraft, training progression, etc. (see examples above). Rolling below an ability value is one way of making achieving success, e.g. a dexterity of '14' on 3d6 – roll below this to succeed. Systems with multiple dice can also look at getting one or more outcomes above a difficulty setting and with all these player characters can have specialisms that add to the sum or allow extra dice.

Experience and training

As a character develops, it leads to progression - improved probability in the performing the various actions in the game. This is either automatic on gaining experience points or by passing training tests. In some, single dice are employed for this, as with the 1d20 and 1d100 (uniform distribution) and this is directly linked to ability to perform a task (combat in particular). In this way, progression through levels is even and in smaller steps. Dice with more 'divisions' allow finer settings - larger sizes and more dice give any scale desired. If 1d20 is insufficient then there's 1d100, 1d200 or even more, as described in Section 18.2.

Other systems involve improvement of the original abilities according to successful completion of various activities in the game. These may in turn affect dice rolls for other actions, e.g. physical power increases and gives an improvement to a combat action. Deftness improves a climbing ability (roll on 1d100). In some cases, the experience gain may not be automatic and a successful dice roll is required. Subsequent rises have a less probable success roll, i.e. it gets more difficult (more reflective of human learning experience (Dennis)).

Combat and melee

Depending on the particular system, the most intensive use of dice is found with combat. In singe die systems, adventures start with low combat action probabilities, but this improves due to general experience and training, plus any bonuses accrued by enhanced character abilities. Some of these systems have been in use for several decades and they have their supporters and critics. The procedure copes with a large variety of bonuses and penalties (in form of +s and -s to the die roll). With systems based on d20, the fighting ability of the protagonists is scaled over the single die according to experience. Thus, a warrior on accruing enough experience points can go from '14' on d20 to '12' (35% to 45%) and by the same proportion on the next level. Using multiple dice such as 3d6, a similar change would give different probabilities (38% to 16%). Probabilities for these are usually 'equal or greater'. For example, an inexperienced warrior could require '17 or more' on 1d20 for a successful strike, P = 4/20 (20%), a top rated one - perhaps '9 or more' = 11/20 (55%). With a system based on 1d100, possibly 'equal or less' is used - an archer needs '45 or less' to hit, P = 45%. In multiple dice systems, the combatants may need to roll below an ability or use their combat dice plus bonuses to overcome opponents' defence and armour, etc. This time, scoring a sum within the established fighting ability can be used - if a space marine fires a pulse gun with ability 12, the marine must get 'a sum 12 or less' on the dice to succeed (P = 74% on 3d6). Whatever the form of the 'attack roll', it can be modified for various effects as above, such as defence and armour of the opponent and special skills, etc.

With a uniform distribution roll, all requirements stated in terms of a single face value are of the same probably, so '83' on 1d100 has P = 1/100 and so has '45'. So, statements like 'that's a difficult action - you need a 89 on the die' is only appropriate if want probability to be 1% as getting '100' has the same probability (similarly for 1d20 but the steps are 5%). To overcome this, events are stated as *inequalities*, but probabilities increase or decrease in a fixed proportion over the die range. Thus, actions that require '77 or more' and '78 or more' on 1d100 have probabilities 24% and 23%, respectively - a very small difference in probability. On d20 '16 or more', '17 or more' have 25% and 20%, respectively. Contrast these with rolling '13 or more' and '14 or more on 3d6 at 26% and 16%, respectively - a much larger difference. On the multiple dice sum scale, divisions from one point on the scale to the next vary and probability falls off rapidly near the extremes. On the other hand, it's not possible to get relatively 'very low' probabilities with single dice -

compare probabilities for 3d6 (consult tables 18.2a, b) (0.5%), 1d20 5%, 1d30 (3%) and 1d100 (1%). However, if you play combat intensive games, or depending on the way the game is played, you may wish 'easier', i.e. more frequent, hit probabilities for combat to speed up play.

In these systems, the average roll (E(X)) can increase with experience, i.e. characters 'get better' at tasks or combat actions, but at the same time the variance may stay the same or indeed, increase as well. Ideally, experience should bring increased ability (higher E(X)) and improved consistency (lower V(X) and SD). In this way, experienced characters have higher probabilities for success and are surer of that success. To-date, there does not appear to be such a system (RPG DESIGN) generally available. Examination of averages and variability of dice sums with different numbers and sizes can produce such a system using the appropriate formulae for single and multiple dice (18.2a/b and 19.1a, respectively):

Spreadsheet Table 20.4

Number of dice	Size	E(X)	SD
1	24	12.5	6.92
2	12	13	4.99
3	8	13.5	3.97
4	6	14	3.42
6	4	15	2.74
8	3	16	2.31
12	2	18	1.73

All these dice groups have the same maximum but an increasing average and a falling SD. Thus, a first or zero level character would start with 1d24 and progress through to 12d2 when highly experienced, confident of a high average with low variability. Of course, this is at the expense of the extra time and complexity with rolling greater numbers of dice.

These points are just some of the many raised in discussion forums on role-play games.

Subsequent dice rolls with other sizes of dice govern damage. Damage for single dice was illustrated above, but it can be based on sums as well. Single dice conveniently give a range of damage levels suitable for cases where you want a maximum roll to have similar probability as those of middling or poor. If medium damage is wanted most of the time with occasional very low or very high levels, then multiple dice are better. Polyhedral and other dice allow coverage of a wide range of damage for different weapons, powers, technologies and types of creatures, races, etc.

The effect of bonuses is accentuated with multiple dice, e.g. + 3 pushes the probability by a very high amount, e.g. compare 3d6 on its own to get '17 or more', $P = 4/216$ (2%) and 3d6 + 3, '14 -18' (16%, 8 times more), with the same event on 1d20 alone, $P = 4/20$ (20%) and 1d20 +3, '14 -20' ($P = 7/20$ (35%) (less than twice).

Attack bonuses accrue due to other factors such as flank (+1) and rear attacks (+2), defence bonuses for dodging, etc. so players try to gain advantage in these way. There are many early and continuing variations of the official rules for such games with dice tables for critical hits, hit location, fumbled attacks, modified by factors for ability skill, defence, armour of opponent, special skills, feats, etc. As imagined, some quite complicated events arise.

20.4 Other dice problems

Famous dice problems
A number of dice conundrums appeared in the past, at which time they took a lot of hard work to elucidate. We can deal with these fairly easily now, with the advantage of current knowledge and computing facilities.

The Newton-Pepys problem (1693; Mosteller[b])
This problem asks the question,

Which is more probable?

'at least one 6 in six rolls'
cf.
'at least two 6s in 12 rolls'
cf.
'at least three in 18 rolls'

Samuel Pepys was unsure of the relative probabilities and asked Isaac Newton to help. This can be answered by use of the binomial function in Excel. Thus, for the first of the three,

P*('at least one 6 in six rolls')* = 1-BINOM.DIST(1-1,6,1/6,1)

$$= 0.665102 \ (67\%)$$

The same function applied to the other two events gives 0.618667 (62%) and 0.597346 (60%), respectively. Therefore, although they are close, the events are listed above in decreasing order of probability. Of, course without the function, workings would require careful calculation (Isaac Newton did them by hand).

Also, by the same function, the *Chevaliers de Méré* (1607-1684) problem (Weisstein) is similar, comparing,

P(*'at least one 6 in 4 throws'*) = 1-BINOM.DIST(1-1,4,1/6,1)

$$= 0.517747 \ (52\%)$$

with,

P(*'at least one double 6 in 24 throws of 2d6'*)

$$= 1\text{-BINOM.DIST}(1\text{-}1,24,1/36,1) = 0.491404 \ (49\%)$$

This poser had contributions from Pascal and Fermat before it was solved. De Méré thought that they should be the same, but trials suggested that the first was more probable (confirmed by the above).

The birthday dice - 'what funny looking dice'

As we now have grasp of some probabilities and have developed tools and some tables, we can end this chapter by applying the dice situation in an unusual way- to solve *the birthday problem*. This is a well-known probability puzzle that confounds some because it counter-intuitive – people tend to disbelieve that the probability is so high for the circumstances. The problem appears in several forms but the main one poses the question:

'How many people need to be present in a group before the probability of at least two of them having a birthday on the same date is greater than 50%?'

Similar questions could be 'what is the probability of two, ten or twenty people in group having the same birthdate?', etc. This puzzle can be tackled in various ways but here we liken the allocation of a person's birthday to the rolling of a 365-sided die ('the birthday die'). We make an assumption that a person's birthday is random and that each year is 365 days long. Within a group of people, take two people - they roll the die and each get a date – one of the 365 days available – it can be the same day or a different day – what is the probability that it is the same?

Remembering our guidance on approach (Section 13.4), we can simplify the problem by scaling down – let's say they use a 6-sided die – basically, we need the probability of getting a double on 2d6 = $^6/_{36}$ = $^1/_6$. For more people, we can increase the number of dice, but this introduces a complication, because now there can be several pairings. With 3 people and 3 dice - '1 with 2', '1 with 3', '2 with 3', etc. for each face value. Table 18.2b comes to the rescue, and using the information, we see that there are 90 single pairs with 3d6, so the probability that 2 are the same is 90/216. However, there are 3 pairs of people ($^3C_2$) and a match could occur with pair 1 or pair 2 or pair 3. Additionally, there is a chance that all three could have it on the same day, so we could broaden the question and say 'At least two matches....'. We would need to add the probability for a match of 3 and so on.

All that's required now is to switch the same reasoning to 365-sided dice and perform the calculations using the methods used to prepare Table 18.2b.

The difficulty is that numbers are looking very large and Excel can't calculate factorials beyond FACT(170). Thus, some functions, like **COMBIN( )** won't work with formula (18.9d) and this also rules out the multinomial route.

However, (18.10) can do this calculation as the large numbers are within the **PERMUT( )** function and,

P('*2 people with same birthday'*)

= PERMUT(365, 2 - 2 +1) *COMBIN(2,2-2+ 0) / 365^2

= 0.002739726 (0.3%)

This is the probability of 2 random people having the same birthday. If we want to calculate for more than three people, we would require values for pairs on 3d365 plus trebles on 3d65. By the time we get to 10 people we would have: pairs on 10d365 + trebles on 10d365 + quads on 10d365 + ... + 'decs' on 10d365, i.e. a lot of computation, bearing in mind that we do not know that 10 people would bring the probability to a level that would answer the original probability question. While such a process is possible by building up a matrix of values and summing them there is an easier way.

We turn to our guidelines again and use the **complement** (Book One Section 4.2.2, formula (4.1)) of the event. We calculate the probability of none of the people having the same day, then subtract this from 1. This will give the figure for at least 2, or 3, or 4 ... up to all 10, 20, 30 .... For this to happen, everyone who rolls the birthday die needs to get a different outcome (birthdate). There's plenty of scope for this as there are 365 of them.

Formula (18.10) will do the calculation by setting 'rpt' =1, or even simpler is the ^sP_n formula and,

$$P = (1 - {}^{365}P_2) = 1- \text{PERMUT}(365,2) /365^2 = 0.00273973$$

This agrees with the calculation above, then for 10 people,

$$P = (1 - {}^{365}P_{10}) = 0.116948 \ (12\%)$$

We see that with 10 people the probability value is still less than 50%.

Further calculation with increasing numbers of people shows that the probability goes over 50% at 23,

P('*at least 2 out of 23 that have the same birthday'*)

$$= 1 - {}^{365}P_{23} / 365^{23} = 0.507297 \ (51\%)$$

If your birthday is today, happy birthday! Using the above, you can calculate the probability of at least one of your friends and relations having their birthday too. Most probability books present the birthday problem, using other methods for illumination in some parts, but all turn to the complementary event for the final stage.

Chapter 21
Cards

21.1 Introduction

Cards and card games are perhaps the most common recreational past-time where the importance of probabilities is recognized. The number of objects is obvious as the 52 cards of the standard 'deck' or pack, although smaller decks and packets can be used. In games, the usual procedure will involve drawing or dealing small numbers of cards (hands) to players or to table. The possible composition of the unseen cards held by the other player or dealer, or the face-down cards on the table are of great interest, as are the probability of getting certain cards and combinations thereof on the deal or by improvement later.

Most card probability problems involve selection of cards from the deck and logically they rely heavily on combinational formulae. The 'beans in a jar' model based on the *hypergeometric distribution* (Book One Section 9.3) is often appropriate for card probabilities, but other distributional functions, namely, the binomial and the multinomial, appear at times.

21.1.1 Card terminology

Cards are identified by the simple notation introduced in Chapter 1 (Book One Section 1.4). 'Faces' refers to the pictures, icons and numbers, etc. displayed on the card when 'turned up'. The term 'picture card' itself refers to kings, queens and jacks (also known as *court cards*). The reverse side of each card, the 'back', must be anonymous and they are patterned with one style to obviate differentiation. Cards have an individual identity, i.e. each of the 52 cards is *distinct*, unless otherwise stipulated. Each card has a *value* of *rank* from 'ace, two, three, etc., up to jack, queen, king', and a *suit* from 'diamonds, hearts, spades and clubs'. Natural subgroups are formed as ranks, e.g. 'queens', as the suits and as colours, e.g. black aces, plus others (e.g. one-eyed jacks). Cards can be specified by name in the complete, full sense with both rank and suit, e.g. 'the 3 of diamonds', be partially named by rank or suit, e.g. 'an ace' or 'a club' or they can be 'unnamed' and take any rank and suit, e.g. 'three kings and any two other cards'.

Abbreviations used for rank names are 'A,2,3 … J,Q,K' ('T' is useful for '10' in some spreadsheet examples) and suits by symbol as ♣♠♦♥. Unnamed cards are indicated by lower case letters, in the style used for dice, e.g. 'a nine with four unnamed cards' = '9,w,x,y,z' and repeats of one of the latter for repeated unspecified rank, e.g. 'x,x' for a pair, 'y,y,y' for a treble, and others such as 'f,f,f' for three face or court cards, etc. Abbreviations are formed with the rank+suit symbol (for a fully named card), e.g. the seven of diamonds and the jack of spades = '7♦,J♠'. Cards in the later layout usually imply order, otherwise for card *content* the style is as 'the 7♦ and the J♠' or '3♥, Q♠ and the 6♦', similar to dice. A fully named card with have the rank first then suit as in the latter examples, a suit symbol or number appearing first would indicate a single card. Thus, '♥,10 '= 'a heart then a ten', and not the 'ten of hearts' the comma separator makes this clear. Repeated members of subgroups can be treated as distinct - all fully named or non-distinct (indistinguishable), e.g. '8♥, K♥ and the 10♥' and '♥♥♥', respectively. Similarly for ranks, with 'J♦,J♠,J♣' and 'J,J,J', respectively.

Probability puzzles can have any number of cards being drawn, but most card games involve dealing out *hands* of two or more cards, e.g. in bridge a hand has 13 cards. There is a multitude of card games, but only a selection is dealt with in detail. Readers will require to consult a card game source (there are many on the web) for guidance on rules, etc.

21.1.2 Shuffling

With card drawing, an assumption is made that a *random draw* is made from a *fully randomised* deck (Book One Section 3.6). *Shuffling* is the method by which this latter state is achieved. Various methods exist and many have been subjected to study in respect of elucidation of the randomisation process and as a route to producing a non-random predictable state for the purpose of conjuring or cheating.

The physical structure of cards is such that when new they can be assumed identical apart from the face markings. Tampering, e.g. crimping or use of marked decks - with back patterns which can be used as indicators, allow identification of one or more cards (Book One Section 3.6). Randomisation can be hindered when, on long use, cards become soiled and sticky and this requires regular replacement in casinos. Assuming that the deck is clean in all respects, the cards can be mixed to a random state by shuffling.

The overhand shuffle

The simplest of shuffles is the overhand method where packets of cards are pulled from the top of the deck onto the opposite hand and stacked, each upon the preceding one. It has the least randomising effect and many of such shuffles would be required to get to an 'acceptable' randomised state with 2,500 times quoted! (Su).

The reader can demonstrate this by a simple experiment: separate the four aces from deck and place them together within the top ten cards. Perform an overhand shuffle by pulling of five packets each time (i.e. approx. ten cards in each packet) – repeat the process. The four aces will have travelled to bottom of the deck, then back up to their original consecutive locations. Any even number of the basic shuffle action will achieve this until slight variations in packet size eventually split the aces (cutting the pack at any point will shift them to the centre region).

Riffle shuffle

The riffle (dovetail) shuffle is perhaps the commonest method and it is based on splitting the deck into two halves and interweaving these together. Done perfectly, the pack returns to its original state after 8 shuffles (Epstein). Such ability is rare and most card players will introduce various degrees of deviation from perfect which allows sufficient randomisation after 7 shuffles (Bayer and Diaconis; Mann).

Many card players will not spend the time doing this and this has been cited as one contributory reason for 'perfect hands' turning up in some deals after a new deck has been brought into a game (Epstein), but see below. Casinos are aware of this and some have introduced another technique called 'stripping' where small packets are pulled from the pack and replaced at other points and now there are card shuffling machines but even these have been criticised. (www.newscientist.com).

Many experiments and trials have been done exploring this phenomenon. The reader can explore a simple experiment by getting a new or sorted deck and giving it one overhand or riffle shuffle and making some of card draws and deals below. Otherwise, for our probability exercises and examples we will assume randomisation to the level of a '7 riffle shuffle'.

Some very large numbers arise in card probability calculations. Interestingly, the number of ways for 52 distinct cards to be arranged enumerates to an enormous number = 52! If you try this in Excel with FACT(52) it will give:

80,658,175,170,943,900,000,000,000,000,000,000,000,000,000,000,000,000,0 00,000,000,000

This is inaccurate due to the limitation of precision (accurate value =

80,658,175,170,943,878,571,660,636,856,403,766,975,289,505,440,883,277,824,0 00,000,000,000).

This means that on shuffling cards a repeat of a particular arrangement is extremely unlikely, i.e., each time you shuffle a different arrangement is produced. Even if you shuffle the cards 100 times a day for 10 years = roughly 400,000 times – this is insignificant cf. the number above. It can be used in some probability calculations such as *'what's the probability of getting an arrangement where all cards are in order of suits and rank?'* = $^1/_{52!}$, or where all of a particular subset are together in the deck, etc. However, these are permutations of the cards, and we can calculate for combinations with smaller numbers (the number of combinations of 52 cards = 1).

21.2 Card probabilities

The calculations here all rely on selections from a 52 card pool, which will be assumed unless stipulated otherwise. As stated above, card problems are analogous to the 'beans in a jar' model, with a mixture of different suit, colour and icon, face values and 'A to K' rank (the beans) in the deck (the jar). Examples with function and formulae for this setting were given in Chapter 17 (Section 17.2). The hypergeometric distribution applies to this circumstance and to the card drawing situation. Many puzzles can be solved using the analogy but it involves viewing various subgroups in the deck. These include suits, ranks, but also others like red and black cards, 1 eyed jacks, court cards, black aces, etc.

The sample space for card probability is based on the deck size, typically 52, for the first action with individual probability (p) of $^1/_{52}$, which applies to single card selections and where cards are replaced. For selection of more cards we can use combination formula ^nC_r to get the sample space, e.g. in *Blackjack* a 2 card hand can be dealt in $^{52}C_2$ ways and in *Canasta*, an 11 card hand can be delivered in $^{108}C_{11}$ ways. Consequently, because of the low base probability, readers will notice that card probabilities can be very low compared with those of coins and dice (some of these instances that follow do not have P as %). The sample space reduces and conditional probability is used when cards are retained, or when changing and replacing cards. In some cases, this probability can change, as with subgroups and when gauging the probability in games after the deal, when new card(s) are drawn and revealed from the remainder of the deck. It should be observed that dealing from the top and drawing from within the deck are equivalent (the 'without replacement' condition) and the same probability applies to each. Thus, dealing three cards from the top has the same probability as spreading the deck out in a fan or a row, face down, and picking three cards. Cutting to a card is equivalent to picking one card by draw or deal, but repeated cuts can be viewed in two ways – if the cut packet is returned and shuffled the next cut(s) are 'with replacment'. If the packet is withheld then the revealed card reduces the sample space and the next cut (lower in the deck) is 'without replacment'.

The 'face down' deck is similar to an enclosed jar of beans – none of the card faces can be seen. Assuming that a random deck has been generated as above, probabilities can be calculated for any selection by dealing or drawing one or more cards. Examples are given for cards as dealt, either as an isolated collection or as actual hands in a game. There will be a number of these but from the viewpoint of individual players, the hand will be viewed as the selection (the hands of other players and relationships are considered later). Thus, in 'four-hand 5-card draw poker', 20 cards are dealt out but probabilities can be investigated for one hand (5 cards) in the first instance.

21.2.1 Card events

As with coins and dice, cards have a number of possible forms of event (Table 21.1). These events follow the calculation structures described in Book One Section 5.3, mainly as *sequences* of cards and card *content*. Card events include selections of single and multiple cards, *fully named* or *partially named cards* or mixtures of these. *Unnamed* cards are alternatives that take on the identity from the remaining available cards according to the constraints of the problem. Events can involve variations in the replacement and order states. As with other objects, the probability for a particular case of a card event applies to all cases of similar nature and this can be used to give 'general case' probabilities. Thus, in Table 21.1, the probability for *'two aces and a 3'* applies to any other pair of card ranks with an odd card and this can be used to get probability for *'any pair'* on a 3-card deal.

Table 21.1 Examples of card probability events

Occurrence	Example ('probability of getting -')
Single cards:	
Deal 1 fully named card	Deal out the 'K♦'
Exchange 1 card to get 1 fully named card	Discard 1 card and draw the '6♥'
Single cut of deck to get 1 partially named	Cut at a red card
Multiple cards with replacement:	
Fully named content	'10♣, 3♦, J♠, 5♣' on a draw of 4, any order
Fully named sequence	'9♦, Q♣, 4♥' on 3 draws, in order
Numerical consec. seq.	'Any J,Q,K' in order [ranks are consec. (content is different)]
Multiple without replacment: (any order unless indicated)	
Fully named content	'2♠, K♦ _and the 8♠_' on deal of 3
Fully named sequence	'9♦,Q♣,4♥' in order on 3
Partially named content	'2 aces and a 3' on draw of 3
Named suits	'Two ♣s' on 2
Named ranks	'Three 9s' on 3
Named +unnamed content	'A pair of 8s' in 5 dealt cards
Hand in *Poker*	A 'flush' in 5 card draw
Hand in *Bridge*	'7 spades' in a 13 card deal
Hand in *Blackjack*	A hand totalling '17' on deal of 2
Exchanges	Improving a pair to two pair in poker
In play *Blackjack*	Getting a third '7' after 'two 7s' on deal
	Getting a face card on hit
Position in deck	All aces together after a shuffle
Higher probability?	'Full house' in *Draw Poker* cf. 'full house' in *Poker Dice*
Expected no. of draws	To pick an ace
Expected no. of deals	To get a prial in *Brag*
Expected occurrence	'blackjack' on 100 deals of 2 cards

21.3 Single card event probability

Single card selections have relatively simple probability calculations. Each fully named card in a standard deck <u>occurs once</u>, and for distinct cards, a full deck is rather like a 52-sided die, and this applies to draws, deals and cuts:

$$P_{single\ card\ distinct} = {}^1/_{no.\ cards\ in\ deck}{}^* = {}^1/_{52}$$

$$P_{single\ card\ of\ suit} = {}^{13}/_{no.\ cards\ in\ deck} = {}^1/_4$$

$$P_{single\ card\ of\ rank} = {}^4/_{no.\ cards\ in\ deck} = {}^1/_{13}$$

$$P_{single\ card\ of\ colour} = {}^{26}/_{no.\ cards\ in\ deck} = {}^1/_2$$

$$P_{single\ card\ of\ court} = {}^{12}/_{no.\ cards\ in\ deck} = {}^3/_{13}$$

... etc.

(for standard deck, but can be other than 52)

In Excel: = 1/52, 1/4, 1/13, ...   *(21.1)*

Thus, '*drawing the king of diamonds ('K♦') from a 52 card deck*':

$$P = {}^1/_{52} = 0.0192308\ (2\%)$$

There is only one 'K♦' in a 'clean' deck, hence the simple calculation. If no further actions occur then the calculation is complete, with a single card selection. This action is equivalent to cutting the pack or a deal of 1 from the top. The same probability applies for any fully named card, e.g.,

$$P('cut\ to\ the\ J\clubsuit)' = {}^1/_{52}$$

Shuffle the cards, deal out top card face up, $P('= Q\spadesuit') = {}^1/_{52}$

In 5-card draw poker, you have four cards '4,5,7 and 8♥' along with 'K♠'. The four hearts contribute to a possible straight flush – you exchange one card to replace the K - what is probability that it will be the '6♥' to complete the straight flush? This draw is based on 47 cards, even though some of these may be in hands of other players – they are all unknown, hence:

$$P('exchange\ 1\ card\ and\ get\ 6\heartsuit\ from\ 47') = {}^1/_{47} = 0.0212766\ (2\%)$$

For **suits and ranks,** the probabilities are adjusted as above according to the subgroup:

$$P('cut\ to\ spade') = {}^1/_4 = 0.25\ (25\%)$$

Alternatively as: $P = {}^{13}C_1 / {}^{52}C_1 = {}^{13}/_{52} = {}^1/_4$

$$P('get\ a\ 10\ on\ drawing\ a\ single\ card') = {}^1/_{13}$$

Or: $P = {}^4C_1 / {}^{52}C_1 = {}^4/_{no.\ cards\ in\ deck} = {}^1/_{13} = 0.0769231\ (8\%)$

$$P('deal\ 1\ card\ or\ single\ cut\ a\ red\ card') = {}^1/_2$$

Other than the card exchange example, further selections will be at the same probability provided drawn cards are <u>returned</u>, followed by shuffling. If shuffling is not performed then the next choice may be biased with even approximate knowledge of where the first card is located.

Single card probabilities can also be solved using distributional function as with one object the replacement issue is not a barrier. The **binomial method** applies, as the necessary conditions are present (successes = 1, no. trials = 1, $p = {}^1/_{52}$ and cumulative = 0 for exact probability). In Excel, for a particular card,

$$P = \text{BINOM.DIST}(1,1,1/52,0) = 0.0192307 = {}^1/_{52}$$

Particular rank: $P = \text{BINOM.DIST}(1,1,4/52,0) = 0.0769231 = {}^1/_{13}$

Particular suit: $P = \text{BINOM.DIST}(1,1,13/52,0) = 0.250000 = {}^1/_4$

The **hypergeometric method** (Book One Section 9.4) is also possible (where, successes(m) =1, trials(n) = 1, objects of interest(M) =1, total no. object(N) =52, 0 for exact P). For the three examples above:

$$P = \text{HYPGEOM.DIST}(1,1,1,52,0) = {}^1/_{52}$$

$$P = \text{HYPGEOM.DIST}(1,1,4,52,0) = {}^1/_{13}$$

$$P = \text{HYPGEOM.DIST}(1,1,13,52,0) = {}^1/_4$$

21.3.1 Expected values for single cards

These distributions also provide the average expected value for a repeated long-term single card selection. The random variable (R(X)) is the number of appearances (count) of a particular card, which can take values 0 or 1 (on the draw), i.e. the event card either appears or it does not. Repeating the experiment requires that each draw must be followed by replacement and shuffling for validity. For example, *'On a single draw with replacement, on average, how many times would I see the Q♥?'* As there is a fixed number of equiprobable independent trials, the binomial distribution formula for the expected value (no. of successes) applies:

$E(X) = n*p$, where $n = 1$, $p = \frac{1}{52}$ and

$E(X)$ '1 draw = $Q♥$' = $1 * \frac{1}{52} = \frac{1}{52}$ (≈ 0)

(also possible using the hypergeometric $E(X) = n * M/N = 1 * \frac{1}{52}$)

Therefore, so the 'Q♥' will appear very infrequently in the long term, but if 52 repeats are enacted, you could expect to see the queen once. With such repeats, R(X) can = 0,1,2, ... 52. The most likely number of appearances (*with replacement* conditions) of the 'Q♥' would be at the binomial mode, in this case once, (P = 0.371457 (37%)).

Similar calculations give the expected value for subgroups,

$E(X)$ '1 draw = 5' = $1 * \frac{4}{52} = \frac{1}{13}$ (≈ 0)

$E(X)$ '1 draw = black' = $1 * \frac{26}{52} = \frac{1}{2}$ (≈ 1)

$E(X)$ '1 draw = ♠' = $1 * \frac{13}{52} = \frac{1}{4}$ (≈ 0)

$E(X)$ '1 draw = court' = $1 * \frac{12}{52}$ (≈ 0)

Again, the most probable 'with replacement' draw or cut is given by the binomial mode:

For rank = 12, 13, P = 0.382697 (38%); for suit = 3, 4, P = 0.421875 (42%).

The single card selections occur in card games and related actions, e.g. cutting the deck to see who deals first (Section 22.2). Simple card games can be built as with betting on single card appearances being specific (e.g. 'a red 3 = a win') or of different rank as with the *'higher or lower'* game (Section 22.4).

21.4 Multiple card event probability

When drawing or dealing two or more cards, some complications surface. Namely, the effect of *replacement* and the *'in order'* vs. *'any order'* states on the resulting probabilities. Card probabilities exhibit a *reducing sample space* (Book One Section 4.2.1) when cards are drawn *without replacement*, which is the case in most games. If drawn card(s) are replaced then the binomial and multinomial distributional functions can be used for some problems (see below).

We do not have the same 'confusability' with card outcomes cf. with those of coins and dice, as all cards are distinct,. Thus, no matter how we arrange a group of cards they will exhibit clearly recognisable orders. The only complication comes when we consider subgroups such as suits or ranks. For example, 'two aces and the 4♣' can be permutated as 'A,A,4♣', 'A,4♣,A', and '4♣,A,A'. We need not concern ourselves with the full identity of the aces in these circumstances, i.e. we can declare that aces are *indistinguishable or non-distinct*, as explained in the terminology above.

Most card games involve card outcome *content* in the assessment of scoring patterns for hands, (e.g. a hand with a meld of 'three 10s' (a set) in rummy), etc. When looking at the deal or draw as it happens, then *sequences* come into consideration, e.g. drawing single cards by player order.

21.4.1 Without replacement

This situation is more typical of actual card games with cards being dealt out in various numbers and retained. As highlighted above, after each single draw, the deck size reduces, as does the total number of outcomes. The formulae used must now take into account that the sample space is reduced by one card after each draw. That is, the events are *not independent*, and the second and subsequent stages are *conditional* on the previous one. The calculation is essentially that of probability formula (5.1b) for the intersection ('AND' case) of *dependent* events (Book One Section 5.2.1). When considering full naming of two or more cards in the event with order, we have:

Card sequence_{without replacment}**:**

$$P = \frac{1}{52} * \frac{1}{51} * \frac{1}{50} \ldots = 1/\,^{52}P_n$$

In Excel: = 1/PERMUT(52,n)

Where, n = no. cards drawn and identified

in order, without replacement

fully named cards   **(21.2a)**

Thus, for '9♦, Q♣, 4♥' in order, n = 3: and,

$$^{52}P_3 = 52!/(52-3)! = 132600 \quad P = \frac{1}{132600} = 0.00000754148$$
$$= \frac{1}{52} * \frac{1}{51} * \frac{1}{50} = 1/PERMUT(52,3) = \frac{1}{132600}$$

This could be the probability for three players being dealt a single one of these 3 cards in order and in fact for any three fully named cards. The event is extremely unlikely but this is a slightly higher probability than the case when the cards are replaced ($P = (\frac{1}{52})^3 = 0.000007112$; Section 21.4.2).

For any order, we have card *content* (combination) *events and* we follow (21.2a) above, but multiply by the number of permutations possible. This is given via the factorial expression (n!), which can be used for two or more distinct items, as are all fully named cards. Thus, we can calculate an outcome *content* i.e. a selection of 2 or more cards without regard to order, on a draw or deal:

Card content combination_{without replacment}*:*

$$P = 1/\,^{52}P_n * n! = 1/\,^{52}C_n$$

In Excel: = 1/COMBIN(52,n)

Where, n = no. cards drawn and identified

any order, without replacement

fully named cards **(21.2b)**

Thus,

P(*'2 black aces, A♣ & A♠ on 2'*)

$= 1/\text{COMBIN}(52,2) = {}^1/_{1326} = 0.000754148$

An example containing more cards, as with 4 cards where all are distinct,

P (*'10♣,3♦,J♠ and the 5♣ as content on deal of 4'*)

$= {}^1/_{52} * {}^1/_{51} * {}^1/_{50} * {}^1/_{49} * 4! = {}^1/_{270725} = 0.00000369379$

$= 1/\text{COMBIN}(52,4) = 0.00000369379$

Similarly, a hand in 4 player rummy (5 cards), on the deal, the probability of *'the 7♠,7♦,7♣ and the A♠,3♦ in any order'*,

$P = 1/\text{COMBIN}(52,5) = 0.000000384769$

This problem appears later in this section on ranks.

Alternatively, using the combination formula, there is one way for 5 cards to be chosen from 5, and this divided by the no. of ways for 5 cards to be selected, *without replacement*, from 52, which confirms the above result:

$P = {}^5C_5 / {}^{52}C_5 = 1/ (52! / (5! * (52-5)!)) = 1/ ((52 * 51 * 50 * 49 * 48)/ 120)$

$= {}^1/_{2598960} = 0.000000384769$

These problems can also be solved using the general ***hypergeometric formula*** (Book One Section 11.3 & Section 17.4), which gives 'any order' outcomes directly. Each individual card is treated as a single coloured ball, giving 52 colours in total. Don't worry, we do not have to allocate all the colours – only the ones chosen. So, if we take three cards - '2♠ *and K♦ and* 8♣', they could be labelled as black, red and green. Choosing these 3 out of 52, for the probability function, all variables, A, a, B, b, and C, c, are 1 or 0,

$$P = \frac{\overset{black}{\binom{1}{1}} * \overset{red}{\binom{1}{1}} * \overset{green}{\binom{1}{1}} * \overset{other}{\binom{1}{0}} * 49}{\binom{52}{3}}$$

$$= {}^{1}/_{22100} = 0.0000452489$$

In Excel, the above becomes:

$$P = (COMBIN(1,1)^{\wedge}n * COMBIN(1,0)^{\wedge}(52-n)) \ / \ COMBIN(52,n)$$

This reduces to the simpler formula of (21.2b), as all the numerator terms become unity.

Any cards, any number, any hand, any game

This latter formula (21.2b) allows us to calculate for any number of distinct named cards in draws more akin to the 'hand' size of various games. It does not matter the size of the hand and the style of dealing provided a random deck is assumed. Thus, the above could be the probability for a hand in 3-card brag. A 5-card hand in draw poker – the probability of getting '2♠,2♦,Q♥,Q♣ *and A♦* 'as the content (combination) of a 5 card hand:

$$P = {}^{1}/_{52} * {}^{1}/_{51} * {}^{1}/_{50} * {}^{1}/_{49} * {}^{1}/_{48} * 5! = 0.000000384769$$
(all such 5 named card hands have the same probability)

By combinatorics: $^5C_5 / {}^{52}C_5 = 0.000000384769$, which is equivalent to the hypergeometric calculation.

Although there are two '2s' and two 'Qs' (a particular 2-pair) in this problem, all cards are distinct as they are fully named. This is also solvable by methods used for ranks and poker hands (see later). This formula can therefore be applied to determine the probability of any hand of cards dealt to you, provided it is viewed as a set of fully named cards. As noticed, probabilities are again extremely low due to all cards being fully named. Similar methods can be used for ranks, suits and other subgroups.

Multiple ranks, suits and other subgroups

Multiple cards from a subgroup such as ranks and suits are accommodated by substituting a count of the number of ways of selecting the cards as partially named rather than fully named, in the formulas above. We consider only the *'without replacement state in any order'* and use a modification of formula (21.2b). For fully named cards, this was $1/{}^{52}C_n$, but for partial named there are more choices and the expression becomes:

Card content for subgroup:

P = (ways to choose r of subgp. on n)/(ways to choose n from deck*)

$$= {}^{sg}C_r / {}^{52}C_n$$

In Excel*: = COMBIN(sg,r_1)/COMBIN(52*,n)*

    * can be other than 52

  Where, sg = unit size of subgroup

    r or r_1= no. of sg cards in draw (r = n)

    n = no. cards drawn and identified

    partially named cards

    any order without replacement   **(21.2c)**

Thus, for *'getting 2 aces on a selection of 2 cards'*,

$$P = {}^4C_2 / {}^{52}C_2 = 6/(26 * 51) = {}^1/_{221} = 0.00452489 \ (0.5\%)$$

The slight complication is due to the selection from the subgroup, which provides the numerator in the calculation. Thus, both counts can be based on combination formula, $^{n}C_{r}$, with the appropriate modification for the subgroup in the formula (21.2c). 'r' must equal 'n' and the formula would require modification for cases where the number of cards selected exceed those named (see below).

Subgroup selection counts (along with Excel functions) require simple substitution of the particular subgroup size:

For rank: sg = 4, r = 1-4, count = $^{4}C_{r}$ = COMBIN(4,r)

suit: sg = 13, r = 1-13, count = $^{13}C_{r}$ = COMBIN(13,r)

'court cards': sg = 12, r = 1-12, count = $^{12}C_{r}$ = COMBIN(12,r)

'black cards': sg = 26, r = 1-26, count = $^{26}C_{r}$ = COMBIN(26,r)

..., etc.

This time, the probabilities are higher as there is more choice. For example, the number of ways of selecting 'two aces from 4' = $^{4}C_{2}$ = 6 and the number of ways of selecting 'seven hearts from 13' = $^{13}C_{7}$ = 1716.

For *both 1 eyed jacks on 2'*, sg =2 and,

$$P = {^{2}C_{2}} / {^{52}C_{2}} = 1/ (26 * 51) = 0.000754148 (0.1\%)$$

Equivalent expressions of probability

Like the multiple selection of fully named cards, subgroup calculations can be performed with other methods, viz. using the *'2 aces on 2'* example to illustrate, equivalent expressions are:

1. *Individual probability*: $P = {^{4}/_{52}} * {^{3}/_{51}} * 2!/2! = {^{1}/_{221}}$

2. *Hypergeometric formula:*

$$\frac{\binom{4}{2}_{\text{aces}} * \binom{48}{0}_{\text{any other cards}}}{\binom{52}{2}}$$

3. *Excel* COMBIN( ) *function*:

    COMBIN(4,2) * COMBIN(48,0)/COMBIN(52,2)

4. *Excel* HYPGEOM.DIST( ) *function*: HYPGEOM.DIST(2,2,4,52,0)

In many of the following examples for card probability, the above methods plus others can be used for calculation but space restrictions limit applying this for all.

Other examples,

P('*4 aces on draw of 4'*)= HYPGEOM.DIST(4,4,4,52,0) = 0.00000369379

P('*Zero aces on draw of* 4')

$$= \text{HYPGEOM.DIST}(0,4,4,52,0) = 0.718737 \ (72\%)$$

So, getting all four aces looks extremely unlikely compared with the much more favourable probability of not getting any.

Mixed subgroups

By extending (21.2.c), we can deal with subgroups in mixtures, thus, for mixed ranks as with,

'*two jacks and three eights on deal of 5 without replacment'*,

$$P = \frac{\binom{4}{2} * \binom{4}{3} * \binom{44}{0}}{\binom{52}{5}} = 0.00000923446$$

This draw is equivalent to a <u>particular</u> 'full house' in poker on a 5-card deal (Section 22.2.1, which confirms the above).

Mixed suits are treated in a similar manner, as for '*getting 3 different suits, a diamond, club and heart on a deal of 3 cards, without replacment'*,

$$P = {}^{13}C_1 * {}^{13}C_1 * {}^{13}C_1 / {}^{52}C_3 = (13^\wedge 3) / (52 * 17 * 50) = 0.0497059 \ (5\ \%)$$

This can be extended for card games to illuminate distribution of suits in the hand, as with *Bridge* and '*5♥s, 3♣s,4♠s and 1♦ in 13 cards'*,

$$P = \frac{\overset{\text{hearts}}{\binom{13}{5}} * \overset{\text{clubs}}{\binom{13}{3}} * \overset{\text{spades}}{\binom{13}{4}} * \overset{\text{diamonds}}{\binom{13}{1}}}{\binom{52}{13}} = 0.00538779 \ (1\%)$$

Specific suit distributions of this nature are much lower in probability than those of 'any four suits' in a bridge hand (Section 22.2.1).

Any rank or suit

To make the selection more general for any rank or suit rather than named ones, requires addition of each number of ways in which such events can occur. For example, '*getting any three of a kind on a draw of 3*' or '*any four cards, all of the same suit on selecting 4'*. We are applying the addition rule ('OR') for these cases, i.e. with the latter example, '*getting 4 of any suit*' can occur in the form of diamonds or as hearts or as spades or as clubs. These events are mutually exclusive – they cannot all happen and we use the union of probabilities formula (Book One Section 5.2.2). We can add up the values or more simply multiply by the number of ways:

Ranks – each rank occurs 13 times, i.e. there are 13 ways to get '3 of a kind', therefore multiply by 13, hence for 'Any three of a kind':

Count = ${}^4C_3$ (aces) +${}^4C_3$ (twos) +${}^4C_3$ (threes) + ... ${}^4C_3$ (queens) + ${}^4C_3$ (kings)

$$= 13 * {}^4C_3 = 13 * 4 = 52 \text{ and,}$$

$$P('\textit{any three of a kind on select 3'}) = 52/\ {}^{52}C_3$$

$$= \text{HYPGEOM.DIST}(3,3,4,52,0) *13 = 0.00235294 \ (0.2\%)$$

Suits occur in 4 forms, thus, 'any suit' events can occur in 4 different ways and we multiply by 4, *'four cards all of the same suit'*,

$$P = \frac{\binom{13}{4} * \binom{39}{0} * 4}{\binom{52}{4}} = 0.0105642$$

$$= \text{HYPGEOM.DIST}(4,4,13,52,0) * 4 = 0.0105642 \ (1\%)$$

This works the other way round as well - recalling the earlier 5 card hand in rummy (Section 21.4), *'the 7♠, 7♦, 7♣ and the A♠, 3♦ in any order'* = 0.000000384769. This can be calculated as *'three 7s, an A and a 3 '* (i.e. any cards that qualify, not the named ones above),

$$P = {}^4C_3 * {}^4C_1 * {}^4C_1 / {}^{52}C_5 = 0.0000246252,$$

but this is for 4 combinations of '7s' and 4 of 'aces' and '3s' (4*4*4 = 64). To get a specific combination, we divide by 64, i.e. only one of these 64 combinations contains the four distinct cards above and, P = 0.000024625/64 = 0.00000384769, which agrees with the earlier figure.

Other applications of the OR operative will be required in card events such as, *'A pair of Aces or Kings'*,

$$\text{No. of ways} = {}^4C_2 \ (\text{aces}) + {}^4C_2 \ (\text{kings}) = 2 * {}^4C_2 = 12$$

and P(*'AA OR KK on select 2'*) = 12/ ${}^{52}C_2$ = 0.00904977 (1%)

Similarly, *'Any 3 hearts or 3 clubs on selecting 3'*,

$$\text{Count} = 2 * {}^{13}C_3 = 572 \text{ and,}$$

$$P(\text{♥,♥,♥ } OR \text{ ♣,♣,♣ } on \ select \ 3') = 572 / {}^{52}C_3 = 0.0258824 \ (3\%)$$

This function can be extended to cover more draws than successes, which is of importance in determining probabilities for card hands in games such as poker (Sections 21.5 & 22.2).

Selections containing both rank and suit

The above events for suits and ranks have no common elements, i.e. it is not possible to get a jack and an eight in one card, or indeed a heart, diamond and a spade at the same time. However, this is not the case if we want mixtures of suit and rank together, in the style *'an ace and a diamond on any single draw'* or *'getting three ♥s and two 7s on dealing 5 cards'*. For these events, there is a common element, viz. in the first example the 'A♦' and in the second the '7♥'.

Suits with ranks

So, in the latter case above, choices (*without replacement*) based on the full complement of hearts and sevens, 13 and 4 respectively, giving a $^{13}C_3 * {}^4C_2$ enumeration, would be incorrect as the '7♥' is common. For these problems, the common card(s) break the conditions of the event. Another example would be *'getting a spade and a jack in two'* – the 'J♠' can be drawn in one – so we need to say *'in 1 or 2 draws'* or *'in exactly 2 draws'*. So, applying this to the ace,

$$P(\text{'Ace and diamond in 1'}) = {}^1/_{52}$$

'Ace and a diamond in 2' can be achieved as:

$$P(\text{'Ace (not diamond) with a diamond (not ace)'})$$
$$= {}^3/_{52} * {}^{12}/_{51} * 2! = 0.0271493 \ (3\%)$$

$$P = \frac{\binom{3}{1} * \binom{12}{1} * \binom{37}{0}}{\binom{52}{2}} = 0.0271493 \ (3\%)$$

aces exc. diamond   diamonds exc. ace

'Three ♥s, two 7s in 5' can be achieved as:

P(*'Three hearts (not 7) with two 7s (not heart)'*)

$= {}^{12}C_3 * {}^3C_2 / {}^{52}C_3 = 0.000253948$

In Excel this given by,

P = 12/52 * 11/51 * 10/50 * 3/49 * 2/48 * FACT(5)/(FACT(3)*FACT(2))

$= 0.000253948$

Other mixtures, such as splitting of suit and rank into further subgroups, are more complex.

21.4.2 Multiple named cards with replacement

The *with replacment* state means that the card deck must be viewed differently and the 52-sided die analogy is apt. Each time a card is drawn, it must be returned so that all 52 are available for the next choice −as with rolling 1d52 two or more times. Some single card cases can be viewed in this manner − as when drawing, with replacment, until a certain card appears (above). The multiple card case 'with replacement' is not usual in card games, but in the interests of exploration, some examples follow. This time, each stage is *independent* as the cards are replaced. The **multiplication rule** ('AND') applies and individual base probabilities can be multiplied together as illustrated in the intersection of events (Book One Section 5.2.1).

Fully named cards:

For a card outcome *sequence*: (permutation):

$P = {}^1\!/_{52}{}^n$ **In Excel: = (1/52)^n** (cf. dice 1/s^n)

Where, n = number of selections with replacement

Thus, for '*9♦,Q♣,4♥*' in order, there are 3 cards and,

$P = {}^1\!/_{52} * {}^1\!/_{52} * {}^1\!/_{52} = ({}^1\!/_{52})^3 = {}^1\!/_{140608} = 0.00000711197$

(Note: 1/PERMUT(52,3) gives the probability for the *without replacement* equivalent (Section 21.4.1))

Essentially, any sequence of fully named cards occurs as <u>one</u> permutation so the numerator will always be 1. The 'any order' state is accommodated by multiplication by the number of arrangements, which in turn depends on whether or not the cards are distinct or non-distinct:

$$P = {}^1/_{52}{}^n * n!$$

In Excel: = (1/52)^n * FACT(n) for distinct

= (1/52)^n * MULTINOMIAL(a_1,b_1,c_1,...) for non-distinct

Where, n = number of selections with replacement

a_1,b_1,c_1,.... = subgroups of distinct and repeated cards

For example, '*the 10♣, the 3♦, the J♠, and the 5♣*' in any order,

$$P = ({}^1/_{52})^4 * 4! = 0.00000014 * 24 = 0.00000328245$$

With repeated cards, as '*K♠, K♠, K♠, K♠ on 4*' where n=4,

$$P = ({}^1/_{52})^4 * 4!/4! = 0.000000136769$$

As seen in the latter case, when a card is replaced it can be drawn again and the formula can be applied to a mix of distinct and repeat occurrences, e.g. '*What is the probability that on drawing 4 cards with replacement, a combination (content) of J♥ with 2♦ and the 6♣ twice is obtained?*'

$$P = {}^1/_{52} * {}^1/_{52} * {}^1/_{52} * {}^1/_{52} * 4!/(2!1!1!) = 0.00000164122$$

For *with replacment* events with multiple presence of rank and suit subgroups, the probabilities are adjusted – treating the repeat cards as indistinguishable:

'*Two clubs on 2*' gives P('♣,♣') = ${}^{13}/_{52} * {}^{13}/_{52} * 2!/2! = {}^1/_{16} = 0.0625000$ (6%)

'*Three face cards on 3*' gives P('f,f,f') = $({}^{12}/_{52})^3 * 3!/3! = 0.0122895$ (1%)

The binomial function is also applicable as cards are replaced. This gives the probability directly as content (a combination). For the above, *'two named aces in 2'* (Section 21.4.1 mixed subgroups *without replacement*),

P = BINOMIAL(2,2,1/52,0) = 0.000369822

(approx. $^1/_{2210,}$ ten times less probable than when retained)

But *'any 2 aces'* has,

P = BINOMIAL(2,2,4/52,0) = 0.00591716

(v. sl. more probable than *without replacement*)

For the *'four K♠s'* above,

P = BINOM.DIST(4,4,1/52,0) = 0.000000136769

Again, note that this is for <u>4 of the same king</u> - 'any 4 kings' can be achieved in the *with replacement* condition as '4 *of the same king'* OR *'3 of the same and 1 different'*, OR ..., etc. This is equivalent to 4 distinct objects being selected *with replacment* in any combination to make up group of 4 - or a d52 rolled 4 times. Here,

No. ways to get 4 'Ks' = 4^4, P = $4^4/52^4$ = 0.0000350128 and,

P(*'any 4 non-distinct'*) = BINOM.DIST(4,4,4/52,0)

The binomial does not apply for the mix of ' *two 6♣s with 2 other named cards'* example above, but the multinomial (Book One Section 11.2) does. Individual card events are A,B,C ..., etc. each with probability = $^1/_{52}$, with each card appearing one or more times, to a total of n (cards drawn) so here, A = 'J♥', B = '2♦' (once each) and C = '6♣' (twice), then:

P = MULTINOMIAL(1,1,2) * (1/52)^1 * (1/52)^1 * (1/52)^2

   = 0.00000164122

(same probability for any two named and one named repeated configuration)

Applying this function to the equivalent of an actual scoring hand in a card game such as *'two 7s and three Qs on 5, with replacement'* (a <u>particular</u> full house in *Poker*),

$$P = \text{MULTINOMIAL}(2,3) * (4/52)\wedge 2 * (4/52)\wedge 3 = 0.0000269329$$

This of course does not exist in poker as it is created by returning the cards. Again an extremely rare event but it is almost three times more probable than the real poker game 'without replacement' equivalent calculated above (Section 21.4.1, Mixed subgroups), where $P = 0.00000923446$.

21.4.3 Expected values for multiple cards

The hypergeometric distribution formulae (Book One Sections 9.4 & 11.3) can be used to calculate parameters for each of the cards in a multiple selection *without replacement*. If the interest lies with 1 card within the selection, the formulae of (9.3a; 1 subgroup) are appropriate otherwise the general distribution formulae (11.2a) are used. For the expected values, these are $E(X) = n * M / N$ and $E(X_i) = n * M_i / N$, respectively.

$M_i = M_1, M_2$ ... is the number of objects in the individual subgroups as distinct from M (one subgroup) for the univariate.

The random variable (X or X_i), signifies the number of appearances of each card(s), on a nominated number of draws. For fully named cards, this variable can take only two values: 0 or 1, it (or they) either appears in the draw or does not. For example, '*On drawing 4 cards without replacement, how many times would the 9♥ appear, on average?*' There are 4 cards drawn (n) and there is but 1 subgroup instance (M) and a total of 52 cards (N), hence,

$$E(X) \text{ '9♥ in 4'} = 4 * {}^1/_{52} = {}^1/_{13} (= 0.076 \approx 0)$$

Calculating $E(X)$ the basic way is by summing of the incidences and their probability, ((8.1a), Book One Chapter 8),

$$E(X) \text{ '9♥ in 4'} = 0 * P(\textit{'not get 9♥ on 4'}) + 1 * P(\textit{'get on 4'}) = {}^1/_{13}$$

This agrees with the distributional calculation and for these cases, if the card(s) appears, it can only be once. Therefore, when drawing 4 cards repeatedly *without replacement*, on average the incidence of the '9♥' would be below one. This can be interpreted as: if you ran a compound experiment of the deal of 4, 13 times, then repeated it over and over, as a binomial experiment (Book One Sections 8.2.2 & 9.2) the card would appear once, on average,

Original $p = 0.076$ and $E(X)_{binomial} = n*p = 13 * \frac{1}{13} = 1$

The other cards in the hand 'not=9♥' have $E(X)$,

$E(X)$ 'other 3 cards' $= 4 * 51 / 52 = 3\frac{12}{13} \approx 4$

Therefore, the others appear approx. 4 times on average i.e. they make up all 4 cards 'more often than not'.

The same expectation applies to any other fully named card. For two fully named, the separate expectation for each is 0.0769231 ($\frac{1}{13}$), etc. Thus, the expected value for fully named card(s) is between 0 and 1, when hand sizes are less than the deck size. The only way to get it to unity is to deal out all cards. The probability of getting the card is now 1, viz.,

$P = 1/(^{52}C_{52}) = 1$

$E(X)$ '9♥ on 52' $= 52 *1 / 52 = 1$ and for others,

$E(X)$ 'not 9♥ on 52' $= 52*51/52 = 51$

Larger hand sizes give larger expectation for fully named cards. Dealing out 13 cards as in a bridge hand would give an expectation of $\frac{1}{4}$ for a particular card. The whole deck is dealt out in this game and as above the expectation for all cards is 1, i.e. they will all be there somewhere within the 4 hands.

If other individual cards, taken together, are specified in a hand, the expected value gets lower e.g., *'What is the long term average for the appearance of the 9♥ along with 3♦, on a 4-card draw?'*

One simple way to get the answer for this is to treat the 2 cards as a subgroup separate from the other cards. The probability of drawing them on a deal of 4 using the 'unnamed' procedure below (Section 21.5) is,

$$P('get\ 2\ on\ 4') = {}^{2}C_{2} * {}^{50}C_{2}/{}^{52}C_{4} = 0.00452489\ (0.5\%)$$

Using the E(X) formula as above,

$$E(X) = 0 * P('not\ get') * 1 * P('get') = 0.00452489 = approx.\ 1/200$$

A much-reduced expectation compared with one card.

For the expected value of '2 on 2', e.g. *'the A♥ with the Q♠ on 2'* in a *Blackjack* deal, the probability (21.1b), and E(X) would be lower,

$$P('A♥,Q♠\ on\ 2') = 1/({}^{52}C_{2}) = 0.000754148\ (0.1\%)$$

$$E(X) = 1 * 0.000754148 = approx.\ 1/1000$$

This time, with no inclusion of additional combinations due to unnamed cards, this hand with 2 fully named cards' appearance would average out much nearer 0 than 1 and once every 1000 deals from a fresh deck.

The above probability information may be useful to know in some circumstances but generally fully named card expectations are low. The card player will be perhaps more interested in a wider assessment, viz. the probability of getting specified subgroups of cards, typically, ranks and suits where probabilities will be higher.

Expected value for multiple cards of a suit, rank and other subgroups

With two or more cards being drawn, in *without replacement* conditions, we can follow the multiple distinct card procedures above for subgroups with the appropriate adjustment of base probabilities. This time the random variable is the number of times the particular subgroup appears. It can take values of 0,1,2,.. etc. up to the value of n, the number of draws, provided this is equal or less than the maximum number possible (e.g. 4 for rank, 13 for suit and 6 for 'red Qs Js Ks', etc.).

Thus, taking <u>suit</u> as the subgroup, '*On drawing or dealing 4 cards without replacement, how many diamonds would appear on average?*' There are 4 cards drawn (n) and there are 13 diamonds (M) and a total of 52 cards (N). The random variable for the number of diamonds can be one of 0,1,2,3, or 4, hence,

$$E(X) \text{ '4 draws, no. } \blacklozenge s' = 4 * {}^{13}/_{52} = {}^{52}/_{52} = 1$$

This agrees with sum of the probabilities and incidence from 0 to 4 diamonds. So, on average, when drawing 4 cards, one diamond would appear. This average would be the same for any other suit. With deals of 4 cards, on average you will get 1 of each suit - try a few deals to see.

The expected value for a <u>particular rank</u> on 4 selections, would be,

$$E(X) \text{ '4 draws, no. 9s' } = 4 * {}^{4}/_{52} = 4/13 = 0.307693$$

Nearer to 0 than 1, or approx. once in every three runs of the 4 draws.

With more possibilities for the random variable, the **mode** of the distribution becomes of more interest. In this case, it gives the most likely number for the diamonds. The mode of the hypergeometric distribution is given by (9.3b) (Book One),

$$v = (n+1)*(M+1)/(N+2) = {}^{5*14}/_{54} = 1.3$$

Hence, mode = 1   (P(*'at mode for suit'*) = 0.438848 (44%))

On such deals of 4 cards, one diamond is the most probable outcome content.

For mixes of ranks or suits, the expected value of the multi-hypergeometric would be required, namely,

$$E(X_i) = n * M_i / N$$

For example, 'Qs' and 'Ks' on deal of a 13 card *Bridge* hand,

$$E(X) \text{ 'Qs on 13' } = 13* {}^{4}/_{52} = 1$$

This applies to the 'Ks' as well, so on average these ranks appear once in a bridge hand (but not necessarily together). For suits in a bridge hand, e.g. clubs,

$$E(X) \text{ 'no. of particular suit on 13 cards'} = 13 * {}^{13}/_{52} = {}^{169}/_{52} = 3.25$$

Therefore, each suit would appear about 3 times (simulated deals confirm this).

Expected occurrence on repeated deals with multiple cards

As with other randomising devices, we can calculate other types of expected values. One is where the original deal is repeated and treated like a binomial experiment, as in the example above. So, on 26 runs of the 4-card draw, on average,

$$E(X) \text{ 'no. app. } 9\heartsuit \text{ on 4, 26 deals'} = 26 * 1/13 = 2$$

A hand with the '$9\heartsuit$' appears twice, and a hand with 'any 9' appears 26* 4/13 = 8 times.

The *Blackjack* deal with 1000 repeats gives,

$$E(X) \text{ 'no. app. } A\heartsuit, Q\spadesuit \text{ on 2, 1000 deals'} = 1000 * 0.000754148 = 0.75$$

$$\text{(approx. once)}$$

This single appearance is for a <u>particular</u> blackjack hand ('$A\heartsuit, Q\spadesuit$'). If we widen the scope and make it for any one of the appropriate subgroups, the expectation will rise. Any ace with any card of value '10' qualifies as a blackjack. The probability for a 'blackjack' when 2 cards are dealt out from a fresh single deck is, P('*blackjack on 2*') ≈5% (Section 22.2.1) and,

$$E(X) \text{ 'no. app. bj. on 2, 1000 deals'} = 1000 * {}^{5}/_{100} = 50$$

Adjusting for a smaller number of deals, say 20, the expected value would be 20 * ${}^{5}/_{100}$ = 1. The winning combination would appear once – this would change as cards are dealt out.

In *Texas Hold 'em*, '*two diamonds on the deal*' have probability,

$$P = {}^{13}/_{52} * {}^{12}/_{51} = 0.0588235 \ (1/17)$$

On 17 repeats of this experiment how often would 2 diamonds appear?

$$E(X) \text{ 'no. app. '$\blacklozenge\blacklozenge$' on 2, 17 deals'} = 17 * {}^1/_{17} = 1$$

The 'two cards suited in diamonds' would appear once on average, within 17 deals (generally $E(X) = n * p$).

Average number of deals required

The other type of expected value is the average number of deals to get a target hand. For this, we use the probability of the multiple event and base the average on assumption of a geometric distribution (Book One Section 10.2). The random variable tales values of 1,2,3, ... ∞, but the average number of deals is given by 1/p, formula (10.1a). For the *Blackjack* problem above, this gives,

$$P('A\heartsuit, Q\spadesuit \text{ on 2'}) = 0.000754148$$

$$E(X) \text{ 'no. deals to get '$A\heartsuit, Q\spadesuit$' on 2'} = 1/0.000754148 = 1326$$

This low probability event involves fully named cards and it will take a long time to appear on average. The most likely deal number is the first (the mode of the distribution),

$$P = \text{NEGBIO.DIST}(0,1,0.000754148,0) = 0.000754148$$

All following probabilities will be lower. Contrast this with the general blackjack, then,

$$E(X) \text{ 'no. deals, blackjack on 2'} = 1/.05 = 20$$

Much fewer deals compared those for a fully named instance above.

As seen, lower probability hands require more deals on average. If the probability of a full house is 0.0014405 and in a session of poker, how many hands would pass, <u>on average</u>, before a full house is dealt?

$$E(X) \text{ 'no. deals full house'} = 1/0.00144057 \approx 694$$

It would need to be a long session, but remember this and the other examples are for the hand <u>on the deal</u> from the individual player's point of view. Much fewer would pass before a pair (P = 0.422569) one of the commonest hands, makes an appearance,

E(X) 'no. deals pair' = 1/0. 422569 ≈ 2

The formulae used so far generate probabilities for subgroups of cards where the draw number matches the requirement, i.e. 'getting 5 of X with 5 draws', etc. For outcomes where the number of cards is not the same as those specified and named, then one or more cards can take a range of identities based on the remaining cards and they are referred to as 'unnamed'.

21.5 Card selections with named cards and unnamed cards

Making selections with a larger number of cards than those named fully or named by subgroup, means that some cards are made up by choices from any of the remainder. The identity of these cards is not specified, hence the description **unnamed cards**. These are not *wild cards*, such as a joker, which can take any suit and rank, but they represent a range of possible outcomes. This is similar to the dice circumstance in a previous chapter. The resulting event can therefore occur in more ways due to the multiple identities for the unnamed cards.

The more well-known of such instances are found with some card hands in the game of *Poker*, where two or more of a rank (e.g. a pair) are combined with one or more unnamed cards. Thus, instead of merely calculating the probability of '2 aces' with a 2 card selection, we would be more likely to be interested in this event for 3-card and 5-card games such as *Brag* and *Poker*, respectively. Abbreviations used for these events are similar to those described above (Section 21.1.1) and are of the style 'Q,Q,x,y,z' (two queens and three unnamed cards) and 'x,x,x,y,z' (a treble rank plus two unnamed), '♣,w,x,y,z' (a club with four unnamed), etc. It is of course possible to have fully named cards along with unnamed ones in events and mixtures of any of these forms. As dealt out for games, such hands do not consider order or replacement, and all events are card combinations as *content*.

To achieve such probabilities requires defining other subgroups. Taking the simplest example of a single fully named card and a single unnamed allocation: *'Getting the jack of clubs on a deal of two cards'*. This could be in the form 'J♣,x' (any order) and identifies one subgroup as the jack and all the remaining cards as the 'other' subgroup. There is only one way to get a fully named card so the calculation is simpler, using individual probability for each stage,

$$P = \frac{1}{52} * \frac{51}{51} * 2! = \frac{2}{52} = \frac{1}{26} = 0.0384615 \ (4\%)$$
$$(\text{or as } \frac{1}{52} * \frac{51}{51} * {}^{2}C_{1} * 1! \text{ - see below})$$

The probability for the 'J♣' is $\frac{1}{52}$ and the choice for the second card is from any one of the remaining 51 cards. This would be for an 'in order' condition, but for 'any order' we could get the jack on the 2nd choice, hence, multiply by 2!. For more unnamed cards, calculation using individual probability requires additional steps as done for dice with direct calculation (formula 18.9b), e.g. 'J♣,x,y,z',

$$P = \frac{1}{52} * \frac{51}{51} * \frac{50}{50} * \frac{49}{49} * {}^{4}C_{1} * 1! = \frac{1}{52} * 4 = \frac{1}{13} = 0.0769231$$
$$= COMBIN(1,1) * COMBIN(51,3)/COMBIN(52,4)$$

(In the individual probability calculation, the named card is 'slotted in' and the arrangements for it are included)

Returning to the single unnamed example, the combinatorial approach can be taken. We view that the 'J♣' can only be selected in one way, ${}^{1}C_{1}$, and the unnamed card must come from the remaining 51 cards. The unnamed card can be any one of these cards, thus there are ${}^{51}C_{1}$ ways of choosing it and the probability is,

$$P(\textit{'J\clubsuit on deal of 2 ['J\clubsuit,x']'}) = {}^{1}C_1 * {}^{51}C_1 / {}^{52}C_2$$

$$= \binom{1}{1} * \binom{51}{1} / \binom{52}{2} = 0.0384615 \ (4\%)$$

This is equivalent to the hypergeometric formula and in Excel:

$$P = \text{HYPGEOM.DIST}(1,2,1,52,0) = 0.0384615, \text{ agreeing with above.}$$

It must be pointed out that this result includes the possibility of another jack appearing and this issue is discussed below. Accepting this for the moment, any number of unnamed can be included with fully named cards:

$$P(\textit{'J\clubsuit on deal of 5 ['J\clubsuit,w,x,y,z']'}) =$$

$$= \text{HYPGEOM.DIST}(1,5,1,52,0) = 0.0961538 \ (10\%)$$

$$P(\textit{'J\clubsuit,5\spadesuit and 9\blacklozenge on deal of 7 ['J\clubsuit,5\spadesuit,9\blacklozenge,w,x,y,z']'})$$

$$= \binom{1}{1}^{3} * \binom{49}{4} / \binom{52}{7} = 0.00158371 \ (0.2\%)$$

This latter example cannot use the Excel function as there are four subgroups selected from the main one.

Changing the above problem to 'getting _a single jack_ on a deal of 2 cards', means that one subgroup would be all the jacks, as we have a partially named instance. Now, the single jack is selected from a possible 4 and the unnamed card from the remaining 48 cards, giving,

$$P(\textit{'a J on deal of 2'}) = \text{HYPGEOM.DIST}(1,2,4,52,0) = 0.144796 \ (14\%)$$

This time, the unnamed card cannot be a 'J' as they are excluded from this selection.

Multiple partially named cards can easily be incorporated, as with 'getting *two nines (a pair) on dealing a hand of 3 cards'*:

$$P = \binom{4}{2} * \binom{48}{1} / \binom{52}{3}$$

$$= 0.0130317 \ (1\%) \ (\equiv \text{HYPGEOM.DIST}(2,3,4,52,0))$$

In this case, the selection from 48 cards cannot include '9s', so only 'a pair of 9s' would result. We can use the hypergeometric function for most of these types of problems, with any number of named cards (including zero) and unnamed ones,

P('*pair of 9s on deal of 5'*)

= HYPGEOM.DIST(2,5,4,52,0) = 0.0399298 (4%)

P('*treble of 9s on deal of 7'*)

= HYPGEOM.DIST(3,7,4,52,0) = 0.00581877 (1%)

P('*zero 9s on deal of 4'*) =

= HYPGEOM.DIST(0,4,4,52,0) = 0.718737 (72%)

This latter example is equivalent to '*any 4 cards except 9s'*, which can also be calculated as,

$$^{48}C_4 / {}^{52}C_4 = 0.718737$$

P('*one 9 + four other cards but not 9s'*)

= HYPGEOM.DIST(1,5,4,52,0) = 0.299474 (30%)

In a similar manner, other subgroups along with unnamed:

P ('*five court + any five other cards'*)

= HYPGEOM.DIST(5,10,12,52,0) = 0.0329419 (3%)

P ('*three spades + any ten other cards'*)

$$= \text{HYPGEOM.DIST}(3,13,13,52,0) = 0.286330 \ (29\%)$$

This is equivalent to a *Bridge* hand with three spades. The other 10 cards can include numbers of other suits ranging from 0 to 10 (see below for more suit examples).

There is a major consequence to use of the hypergeometric function in this way because, like its 'with replacement' equivalent, the binomial in the dice case, it allows the unnamed cards to take any of the remaining card identities. Hence, in the *'a pair of 9s on a deal of 5'* above, other pairs, and trebles are possible, e.g. a content of *'two 9s, two 6s, and a J'* or *'two 9s and three Qs'* could occur. This is similar to the situation with dice (Section 18.3.4), where the binomial was used. If we are happy with such outcomes to be included then we can use combinational formulae or the function as above.

21.5.1 Constraints for unnamed cards

A more common situation is that such circumstances would infringe the requirements of the event, in that we require the unnamed card(s) to be different in rank or suit from the specified ones <u>and</u> to be free from repetition, etc. Thus, *constraints* are in force, and in the example *'a pair of 9s on deal of 5'* the objective is '9,9,x,y,z' where, $x \neq y \neq z \neq$ '9'. So, applying the combinational selection of 3 from 48, $\binom{48}{3}$ would over-estimate the count. To overcome this hurdle, we need to *partition* the selection by first selecting 3 ranks from the remaining available ranks, then selecting 1 card from each:

No. of ways to select 3 ranks from 12 = $^{12}C_3$, then,

No. of ways select 1 from each of 4 = $^4C_1 * {}^4C_1 * {}^4C_1$

To follow this calculation, first view the ranks as subsets - there are 13 in a 52 card deck – one has been used to get the '9s' – so the '9' subgroup cannot come into the next calculation and there are 12 available for the 1st choice. Thus, the initial part of the next stage of selection is $^{12}C_3$. Remember that all 12 subsets of ranks are distinct so using the combination formula will select 3 different ranks, thus no more pairs or trebles can occur. Each of the chosen ranks must be multiplied by the number of ways to choose 1 from the 4 cards in the selected rank = $^4C_1 * {}^4C_1 * {}^4C_1$ which will give 3 distinct rank cards. These two components are multiplied together and divided by the total ways of selecting 5 cards from the deck ($^{52}C_5$) to give,

P('*two 9s plus 3 distinct non-9 cards*')

$= (^4C_2) * (^{12}C_3 * {}^4C_1 * {}^4C_1 * {}^4C_1) / (^{52}C_5) = 0.0325053$ (3%)

This calculation cannot be performed by the hypergeometric function, as there are more than two groups involved, and it requires the **multi-hypergeometric formula** ((11.2b) Book One Section 11.3),

$$P = \binom{4}{2} * \binom{12}{3} * \binom{4}{1}^3 / \binom{52}{5} = 0.0325053 \text{ (3\%)}$$

We can easily broaden the scope of the method for more named cards, thus, '*getting 3 jacks in a 5 card deal*' using the Excel equivalent of the above style,

P = COMBIN(4,3) * COMBIN(12,2) * COMBIN(4,1)^2 / COMBIN(52,5)

= 0.00162526

General repeated subgroups

Extending these examples to the probability of a more general 'pair' or more of a subgroup in a hand of cards requires application of the addition rule ('OR') as described above for 'any rank' type problems. Thus, using the 'pair of 9s' result,

P('*Any rank pair on deal of 5*') = P('*a particular pair*') * 13

= P(pair As) + P(pair 2s) ... + P(pair Qs) + P(pair Ks)

= 0.0325053 * 13 = 0.422569 (42%)

Thus, provided we have the probability value for a particular instance of the subgroup we can easily calculate the general version:

P('*Any treble on deal of 5*') = P('a particular treble') * 13

= 0.00162526 * 13 = 0.0211285 (2%)

P('*Any 4 of a kind on 7*') = P('a particular 4 of a kind') * 13

P('*7 of any suit on 13*') = P('7 of a particular suit') * 4

P('*9 of any one colour on 10*') = P('9 of a particular colour') * 2

P('*Any of a subgroup on ...*') = P('... of a particular ...') * ..., etc.

This concept can be carried through for any other choices of hand composition for games, with the multi-hypergeometric formula perhaps giving the best visualization, (further examples appear in Section 22.2 and in Problems & Exercises) but complications occur as the number of unnamed cards increases (see below).

E(X) of card hands in games

Expectations for hands on the deal relate to their occurrence in games sessions and to the possible return in betting card games. We have looked at expected occurrence of hands on the deal above. For later in the game, where a choice is made on exchanges in card games there may be more than one scoring outcome possible. The expected value (typically as an expected return) would give a single value for the exchange instead of a list of individual event probabilities (see *Poker* games Section 22.2.3)

Chapter 22
Card games and card problems

22.1 Introduction

The cards have been dealt and you are looking at your hand. How can probability help? There are of course many card games, with many rule variations and embellishments, each reader will no doubt have their favourites. Only a sprinkling of some familiar examples is found below. We can simplify the approach, as there are some common probability issues:

- o   Initial deal probability
- o   Probability of other players hands
- o   Probability of improvement of the hand
- o   Probability and play of the hand

First is the probability of getting particular combinations of cards that signify some scoring value in the game, on the initial deal out. These starting hands may or may not decide at the onset which player has the winning hand. Usually, the higher the value of the hand, the higher the chance of winning. This applies for many card games, e.g. getting all the spades in a 13 card bridge hand, getting three red threes in a canasta, or getting four aces in *Poker* are all possible 'first deal' hands, although they are very rare occurrences. The probabilities for these hands are calculated with the classical formula using the ways to get such a hand divided by the ways to deal that number of cards in one hand. The probability along with other factors, allows an immediate assessment of the value of the hand in the game.

After the deal out of cards the probability of other players' hands is also of interest – what is the probability that your opponent(s) have a better hand than you do? Thirdly, comes the improvement of the hand by replacing and exchanging cards and in some games this follows on into the play. Here, choices can be made and probability can give guidance. Also important are certain other activities - your position of play in relation to other players, bidding or betting actions, the number and identity of cards they choose or exchange, etc. Probability based decisions are augmented by these 'signals' as to appropriate play, complicated in some games in that players can bluff their true intentions.

A good memory is an advantage in many games, including cards. If a count is kept of cards that are known, even at the simplest level, then probability-based judgments are improved. For the games below, a brief strategy guide is indicated but only in so far as probability enters the decision making process. All the card games in this book are available in video format so they can be played on the table top at home, in casinos, and clubs, or on personal computers, consoles, tablets and smart phones, etc.

22.2 Probability with card games

High card cut or high card draw

At the start of many card games, a common action is 'to cut for deal' – is the probability 50:50? This problem was mentioned in Section 15.2.1, but it does not appear to be concisely analysed for the card case in probability texts. A single cut is like a deal of one card, but the cut(s) for other player(s) can be done under different circumstances as with shuffling between cuts, etc. The simplest way is to view the cuts as two cards dealt in order (2 players), treated as the 'without replacement' state (if the first packet is replaced then 'with replacement'). Using the procedure similar to that described for dice in Chapter 15, we can count the instances of losing for player 2 (Spreadsheet Table 22.1).

The count for the ranks goes from where player 2 can lose in 12 ways against an 'ace' (player 1) down to 1 way for against a '3'. So, using (15.0),

Count = 12 + 11 + ... 2 + 1 = 12*(12+1)/2 = 78

Spreadsheet Table 22.1

1st cut (card) Player 1	2nd cut (card) Player 2	Lose (p2)
A	K,Q,J,10,9, ... 2	12 ways * 4 * 4
K	Q, J,10,9... 2	11 ways * 4 * 4
...	...	...
3	2	1 way
	Ways in total =	1248

Each of these losing ranks occur in 4 suits, then these can be paired with 4 different aces,

Count = 78 *16 =1248

These are permutations for 'player1 then player 2' and in this form, 2 cards can be dealt from the deck in,

PERMUT(52,2) ways = 2652, then,

$$P('Player\ 2\ loses') = {}^{1248}/_{2652} = 0.470588\ (47\%)$$

The same probability applies for player 2 on the second cut. Therefore, the chances are not 50:50 as there is a chance of a tie, requiring further trials to resolve. These can be ignored and further cuts enacted and we get the probability by summing the geometric series where,

Common ratio = tie probability = any pair on 2 cards

= HYPGEOM.DIST (2,2,4,52,0) * 13 = 0.0588235

$$P('A\ wins') = 0.470588\ /\ (1 - 0.0588235) = 0.5\ (50\%)$$

An alternative method is to calculate the probability of a tie, then subtract this from 1 and divide by 2. Leading on from high card cuts, we look at a game that exemplifies probability use in a relatively simple manner.

Higher or lower

This game has at least one other name (e.g. *Hi-lo*). It prompts the player to make a probability estimate before a simple choice is made. One or more cards are laid out face up from a randomised deck and players can bet points or chips on whether the next card is higher/lower/same in rank or suit / or colour, etc. giving a variety of possibilities. Generally, probabilities are calculated as,

P = no. of cards different or same, etc. / no. of cards remaining

Those after the first card is revealed are the simplest to calculate. For example, one card is laid out to start, the '9♣' - 51 cards remain. Probability calculations require counting the number of cards, which fit the wager (Table 22.1).

Table 22.1 *'Higher or lower'* initial odds for 1st card = '9♣'

Outcome	Qualifying	No.	P	Odds in favour
Lower	8s,7s,...2s	28	28/51	1:1
Higher	10s, Js, Q, ... As	20	20/51	2:5
Same rank	9s	3	3/51	1:17
Different colour	reds	26	26/51	1:1
Same colour	blacks	25	25/51	1:1
Different suit	♠s, ♦s, ♥s	39	39/51	4:5
Same suit	clubs	12	12/51	1:4

As more cards are laid out either during play or on the deal, probabilities will change from these (or the deck could be shuffled and a new card drawn to make the odds constant). Players can score points or chips on bets and odds can be set up. A player who wins one card can continue, letting the bet ride, otherwise the play rotates. However, this introduces a jeopardy element into the game as the whole bet could be lost. Good play relies on retaining counts of past cards and adjusting the odds accordingly.

The disadvantage is that eventually some bets will become void in that the probability will be certain, or very low, assuming a one-deck game, played out until the cards are spent. Thus, if three aces are out already, the probability of getting a card the same (the 4th ace) is very low, but the game is helpful in illustrating probability use for beginners and can be played with a smaller deck for different odds.

22.2.1 Probabilities for hands on the deal

Moving onto card games with a structure of 'deal - exchange - play, we begin with the deal. Initial probabilities for the hand structure of one player can be calculated relatively easily. These initial values provide us with a picture of the potential of your current hand and an estimate of opponents' (Table 22.2). For convenience, all calculations are shown using the ^nC_r formula style.

Poker card games

The probabilities of card combinations for the various deal patterns in **Poker** have perhaps received more attention than those of any other card game. At the start of **Draw Poker**, five cards are dealt out face down to two or more players. The probability of the composition of any individual players hand can be calculated using the procedures described above (Section 21.5.1), many relying on the multi-hypergeometric formula (11.2b). The method was illustrated with the example of *'getting any pair with 5 cards'* (*one pair*). A 'single pair' is one of the lower ranked hands. Using the general probabilities in the table, we can get particular cases of interest. For example, many poker venues specify 'Jacks or higher pairs' to stay in the game. A particular pair would be the above general probability divided by 13, then those for 'Js or higher' (4) are added:

$$P = J+Q+K+A = 4 * 0.42/13 = 0.130021 \ (13\%)$$

The calculation method for a pair can be expanded for the other hands in the game, as summarised in the table, giving relatively simple calculations for *three* and *four of a kind* (P = 0.0002401).

Two pairs can be selected from two sets of ranks, with one odd card from the remaining 11 ranks. A *full house* has a joint probability for a treble and a pair. In Section 21.4.1, probability for a particular hand containing *'two Js and three 8s'* was calculated (P = 0.00000923); if we divide the current full house result by (12*13), we get this.

Table 22.2 Initial hand probabilities for some card games

Initial hand	Calculation	P
5-card draw and video poker:		
Pair	$13 * (^4C_2) * (^{12}C_3 * (^4C_1)^3) / {}^{52}C_5$	0.422569 (42%)
Two pair	$^{13}C_2 * {}^4C_2 * {}^4C_2 * {}^4C_1 * 11 / {}^{52}C_5$	0.0475390 (5%)
Straight	$((^4C_1)^5 * 10 - 40) / {}^{52}C_5$ exc. straight flushes	0.00392465 (0.4%)
3oak	$^4C_3 * 13 * {}^4C_1 * {}^4C_1 * {}^{12}C_2 / {}^{52}C_5$	0.0211285 (2%)
Full house	$^4C_3 * 13 * {}^4C_2 * 12 / {}^{52}C_5$	0.00144057 (0.1%)
Flush	$(^{13}C_5 * 4 - 40) / {}^{52}C_5$	0.00196540 (0.2%)
7 card stud:		
Three pair	$^{13}C_3 * (^4C_2)^3 * {}^{10}C_1 * {}^4C_1 / {}^{52}C_7$	0.0184703 (2%)
3oak	$(13 * {}^4C_3 * {}^{12}C_4 * (^4C_1)^4 - 127800) / {}^{52}C_7$	0.0482987 (5%)
Texas Hold 'em:		
Pocket pair	$13 * (^4C_2) / {}^{52}C_2$	0.0588235 (6%)
Two suited cards	$4 * {}^{13}C_2 / {}^{52}C_5$	0.235294 (24%)
Rummy games:		
7card - 3 card set	$(13 * {}^4C_3 * {}^{48}C_4 - 600) / {}^{52}C_7$	0.07539704 (8%)
Canasta - 3 red 3s	$^4C_3 * {}^{104}C_8 / {}^{108}C_{11}$	0.002986 (0.3%)
Whist games:		
7-card - AKQ suited	$1 * {}^{39}C_4 / {}^{52}C_7$	0.000614802
Bridge - 13 any one suit	$(^{13}C_{13}) / (^{52}C_{13})$	0.00000000000157
4♦-4♥-3♣-2♠	$(^{13}C_4 * {}^{13}C_4 * {}^{13}C_3 * {}^{13}C_2) / (^{52}C_{13})$	0.0179593 (2%)
10 point hand (mix A,K,Q,Js)	see text	0.0940511 (9%)
Blackjack:		
Player blackjack	$^4C_1 * {}^{16}C_1 / {}^{52}C_2$	0.0482655 (5%)
Player '17'	$^{16}C_1 * {}^4C_1 / {}^{52}C_2$ OR ${}^4C_1 * {}^4C_1 / {}^{52}C_2$ OR ${}^4C_1 * {}^4C_1 / {}^{52}C_2$	0.0723981 (7%)

A *straight* requires a bit of thought – there are four sets of 13 cards but a straight can have any of the suits. Additionally, the cards must be in numerical order. The number of orders possible per suit from 2 to ace taking

5 at a time can be obtained by counting out, or by use of the formula (18.7c) for dice sequences (Section 18.3.3), substituting the number of card ranks for dice size and hand size for dice number:

No. card seq. = (no. ranks – hand size) +1 = 13 – 5 +1

= 9 sequences with 13 cards

This count requires plus 1, as ace can be high or low giving 10 in total, which is applied to the choice of 1 from 5 ranks (this includes straight flushes so these are extracted).

In the case of a **straight flush,** the choice is restricted as for a straight, but it is also limited to a suit. So, for a straight flush in hearts, instead of having a selection of 1 from 4 for each rank, we are limited to $^1C_1 = 1$; this applies to each card times the number of sequences possible with 13 hearts (10) and multiplied by 4 for all suits = 40. This includes royal flushes (4) and these are subtracted to give the probability for the standard or common straight flush:

$$P = (^1C_1 * {}^1C_1 * {}^1C_1 * {}^1C_1 * {}^1C_1 *10 * 4 - 4)/ {}^{52}C_5$$
$$= 36/{}^{52}C_5 = 0.0000138517$$

We can easily lay out all the **royal flush** hands as only 4 of the 40 possible straight flushes have 'Ace high':

$$P = 4/{}^{52}C_5 = 0.00000153908$$

For **high card** hands, it does not appear to be possible to get the probability directly and the subtraction of some of the other hand outcomes is necessary. The simplest way is by subtraction of all other counts (= 1296460) from the total number of 5 card hands ($^{52}C_5$),

$$P = (^{52}C_5 - 1296460)/ {}^{52}C_5 = 1302500/{}^{52}C_5 = 0.510116 \ (51\%)$$

Alternatively, we can calculate selection of 5 different rank cards, thereby excluding repeated rank incidence (pairs, trebles, quads, full house), then subtract the nos. for flush, straight, etc.

Thus, 'high card' accounts for most ($\approx$ 51%) of the dealt hands in draw poker and over 1 million, out of $\approx$ 2.6 million, are possible. These are made of various specific 'high card' versions, such as 'ace high' and '10 high', etc. The lowest 'high card' hand would be '7 high' as it is the lowest that allows 4 other non-straight cards (a '6 high' and '5 high' would be straights), so we can have 8 different 'high card' forms. Specific 'high card' probability can be achieved by using the above enumeration for the number of suits to be included. This total is modified by subtraction as above including the general 'high card' count and amendment for number of ranks and sequences.

Comparison of the above 5-card draw hand probabilities with the corresponding patterns in *Poker Dice* (5d6), show that all are higher for the latter (excepting 'high card').

Similar approaches to probability calculations are used for other forms of poker but are now based on different hand sizes and for some, the working gets more protracted. For most, the scoring patterns are still based on 5 cards and probability calculations for these versions are similar. More cards in the hand mean that certain hands are more probable (than 5-card poker). Probabilities for all hands (when all cards are dealt out) in *7-card Stud,* except 'high card' are higher than those of *5-card Draw.* The 7-card version also allows for 3 pairs. This is calculated by selecting the 3 ranks for the pairs and taking 2 from each, then selecting 1 from the remaining 10 ranks and picking 1 card,

$$P = {}^{13}C_3 * ({}^4C_2)^3 * {}^{10}C_1 * {}^4C_1 = 0.0184703 \ (2\%)$$

(this hand is counted along with the 'two pair' sets)

Calculations for other hands in 7-card stud become increasingly complicated (Alspach) for workings. As an example, take the probability of 'three of a kind' (relatively simple to calculate in draw poker). For 7-card stud, a first step is to calculate for *'3oak with any other 4 cards',*

$$P = 13 * {}^4C_3 * {}^{48}C_4 / {}^{52}C_7 = 0.0756303$$

However, this gives an inflated probability for 3oak as it can include several other scoring patterns. To limit the possibility of other trebles and quads we need to select 4 other ranks and select 1 card from each, to get '3oak with 4 other different ranks',

$$P = 13 * {}^4C_3 * {}^{12}C_4 * ({}^4C_1)^4 / {}^{52}C_7 = 0.0492471$$

This still includes 4 card combinations linking up with the rank or suit within the treble to give 5 card straights, flushes and straight flushes. These can be calculated separately by a rather tortuous process of counting to produce:

<u>No. straights</u> =

50 (no ways for all treble rank to form 5 straights)

* 4 (each treble in 4 forms)

* 4^4 (each 4 card straights possible occur in 4^4 forms)

= 50 * 4^3 = 51200

<u>No. flushes</u> (including straight flush) =

${}^{12}C_4$ (each 4-card flush occurs in 495 forms)

* 3 (3 suits in each treble)

* 4 * 13 (each suit and each treble)

= 495 * 3 * 52 = 77200

<u>No. straight flushes</u> =

50(no ways for treble suits to form a 5 flush)

*4 * 3 (4 sets of 3 suits in trebles)

= 50 * 12 = 600

These figures are put together and,

$$P = (13 * 4 * 495 * 4^4) - 51200 - (77200-600) / {}^{52}C_7 = 0.04829 \ (5\%)$$

A simpler, but perhaps less comprehensible formula in terms of the adjustment above, is given by the above reference (Alspach):

$$P = (^{13}C_5 - 10) * {}^5C_1 * {}^4C_3 * (({}^4C_1)^4 - 3) / {}^{52}C_7$$
$$= 0.04829 \ (5\%; \text{ cf. } 2\% \text{ for draw poker})$$

The most probable hand on the deal in 7-card stud is 'one pair' (44%) (high card is 17%).

With games with fewer cards, probability usually gets less and calculations get simpler, e.g. a pair with 3 cards in *Brag* or *3-card poker*:

$$P('pair \ on \ 3') = {}^{13}C_1 * {}^4C_2 * {}^{48}C_1 / {}^{52}C_3$$
$$= 0.169412 \ (17\%; \text{ cf. } 42\% \text{ with 5 card})$$

The former game has different hand names for some card combinations,

$$P('running \ flush' \text{ (straight flush)})$$
$$= 12 * {}^1C_1{}^3 * 4 \ /{}^{52}C_3 = 48/{}^{52}C_3 = 0.00217195 \ (0.2\%)$$

$$P('prial('3oak')) = {}^4C_3 * 13 \ {}^{52}C_3 = 52/ \ {}^{52}C_3 = 0.00235294 \ (0.2\%)$$

If you subtract the highest ranking hand '333' (P = 0.000180995), we get the probability for the other prials,

$$P = 0.00217195 \ (0.2\%)$$

Thus, in *Brag*, ordinary 'prials' and the 'running flush' have the same probability values (0.2%), although the prials rank higher (Garibaldi). The expectation (Section 21.4.3) of the top prial is = 1/ 0.000180995 = 5525 deals, so it is about 10 times less frequent than the other prials at approx. 460. Flush and 'run' hands require adjustment for presence of straights and flushes, respectively. When choosing 1 of each rank in the 3 card set there are 4^3 ways of choosing 3 cards for each form of run but 4 will form a flush,

$$P('3 \ card \ flush') = (^{13}C_3 - 12) * 4 / \ {}^{52}C_3 = 0.0495928 \ (5\%)$$

There are more possibilities for a run cf. those in 5-card hands. Using the expression above, the no. of 3 card runs = (13-3) +1 and +1 extra for the ace

= 12. On choosing 3 cards from 13 ranks of 1 suit - 12 form 3 card straights,

$$P(\text{'3 card run'}) = (^4C_1 {}^\wedge 3 - 4) * 12 / {}^{52}C_3 = 0.0325792 \ (3\%)$$

('A,2,3' is highest run in *Brag*)

As seen some of these probability rankings are different to those for *Poker* with 3 cards and from some in 5-card poker, where a flush (0.2 %) is less probable than a straight (0.4%) (Table 22.2).

In *Texas Hold 'em*, a pair on the deal has probability,

$$P = 13 * {}^4C_2 / {}^{52}C_2 = 0.0588235 \ (6\%)$$

In contrast to *Draw* and *Stud Poker*, the most probable hand on the deal in the above game is 'two suited' cards,

$$P = {}^{13}C_2 * 4 / {}^{52}C_2 = 0.235294 \ (24\%)$$

Ultimately, in this game and *7-card stud*, the final hand is made with a choice of 5 from 7 cards but earlier assessment is required as cards are revealed - players are required to make decisions on whether or not to stay in, call, raise or fold, etc.

Generally, the hierarchy of poker hands is according to probability of occurrence on a random deal - the lower the probability, the higher the value, ranging from 'High Card' (lowest) to 'Royal Flush' (highest). The most common 'playable' hand is a pair – the second most probable in draw poker. Generally, the higher your hand is in the ranking, the better your chances are of winning.

Other card game initial hands

Rummy games

Rummy games have similar sought patterns (melds) to poker hands consisting of 'sets' (3oak or 4oak) and 3 or more card straight flushes ('runs'). Initial hand sizes include 5 to 7 cards in *straight rummy*, 10 cards in *Gin* and go to as many as 15 in 2-player *Canasta*. If the rules allow the first stock card to be revealed, the player who has access to this has 8, 11, or 16 cards, respectively. The melds can be suitable for lay down as they are, or can be built on according to the particular game rules. Some sets of scoring cards can be dealt on the deal and parts of the probability calculations are similar or the same as those above. A 3-card set (3oak) in *7-card rummy* follows the same path as that for *7-card stud,* but with rummy ordinary straights or flushes don't count as it's only *straight flushes* that are valid, so we simply subtract the 5-card versions of these as calculated above. For '*3oak with <u>any</u> other 4 cards but excluding 5 card straight flushes* ',

$$P = (13 * {}^4C_3 * {}^{48}C_4 - 600)/ {}^{52}C_7 = 0.07539704 \ (8\%)$$

This still includes 3- and 4-card version straight flushes that take much more figuring out (not shown) but a simulation of the 7-card deal estimates 8% for a single treble with no straight flushes and no 4oak.

If the initiating player gets access to the first card of the discards, probabilities are enhanced,

$$P = (13 * {}^4C_3 * {}^{48}C_5 - 600)/ {}^{52}C_8 = 0.118278 \ (12\%)$$

This illustrates the advantage possessed by this player, which could be viewed as unfair (Scarne).

In *Gin rummy,* hands are based on 10 cards for two players on the deal, thus there are ${}^{52}C_{10}$ possible hands or ${}^{52}C_{11}$ for the non-dealer. Initial probabilities for *Gin* are also complicated and there are more odd cards. To get the number of 3oak sets possible, a rank is selected and 3 cards chosen from it, with 52 triples possible as for straight rummy. From here, the calculation can go in different ways but if we take the simplest first, namely, that the other 7 cards are a random mix of the remaining 48 (after the exclusion of the selected rank), then,

$$P('3 \ of \ a \ kind, \ 10 \ card') = {}^{13}C_1 * {}^4C_3 * {}^{48}C_7 / {}^{52}C_{10} = 0.24202 \ (24\%)$$

Like the above, this value is compromised by inclusion of other trebles, quads and straight flushes, although these are of low frequency. Calculations that exclude these are difficult to do directly but a simulation count gives an estimate of 3oak on its own,

$$P('estimate \text{ } 3 \text{ } \textbf{of a kind} \text{ } in \text{ } 10') = 23\%$$

Straight flush calculations are more complex with 10 cards, but simulation puts the probability of a single straight flush to 3% (the majority (90%) of length 3). To get any 3-card meld is estimated at 26% by simulation.

Going by the above figures, the chances of a complete meld in your rummy hand on the deal are not too low, but you are more likely to have paired ranks and 2-card straight flushes. As imagined (without calculation), the probability of being dealt a 'laydown' Gin hand is extremely remote and the majority of rummy hands require to be augmented during play. The presence of a straight flush, particularly if open–ended and near the middle (6-7-8) is more valuable than 3oak (see below).

More cards are in play in *Canasta*, with 2 decks with 2 jokers each giving 108 in total. Deals are of 11 to 15 cards depending on the number of players. One scoring set is special - 'three red threes' and it has a low probability of appearing on the deal,

$$P('3 \text{ } red \text{ } 3s + \underline{any} \text{ } other \text{ } 8 \text{ } on \text{ } 11') = {}^{4}C_3 * {}^{104}C_8 / {}^{108}C_{11}$$

$$= 0.00298651 \text{ } (0.29\%) = \text{HYPGEOM.DIST}(3,11,4,108,0)$$

The odds improve if you calculate for 2 players in a partnership, i.e. for 22 cards.

Whist and Bridge

With whist games, repeats of a higher rank are valuable and the *flush* pattern is of special interest. Generally, the more of these you have on the deal the higher value your hand. In a simple **whist** game consisting of 7-card hands, we can calculate the odds for getting various mixtures of the suits, using the formulae in Section 21.5.1:

P('*7 cards all of 1 suit on 7'*) = 4 * $^{13}C_7$ / $^{52}C_7$ = 0.0000513064, or as,

P('*7-0-0-0'*) = 4 * HYPGEOM.DIST(7,7,13,52,0) = 0.0000513064

Calculation for mixed suits can be applied as below and this increases probability,

P('*2-2-2-1'*) = 0.0461128 (5%)

All of one suit that matches the upturned trump card is also very rare,

P = $^{12}C_7$ / $^{52}C_7$ = 0.00000591997

For example, this could be a hand where spades are trump, 1 spade is out at the start. The hand consists of 7 other spades selected from the 12 available.

A hand with 3 of trump and any other 4 is much more likely,

P = $^{12}C_3$ * $^{39}C_4$ / $^{52}C_7$ = 0.135256 (14%)

High cards such as 'A', 'K' and 'Q' in a <u>particular suit</u> in a 7-card whist hand, brings the probability down again, even with any other 4 cards,

P = 1 * $^{39}C_4$/$^{52}C_7$ = 0.000614802

Expanding to 'any suit' boosts this by a factor of 4 (P = 0.00245921) but its not until 'any AKQ' is specified that we get above rare levels of probability,

P = 0.0393473 (4%) (These hands could include other high cards)

Bridge hands are also evaluated on the basis of distribution of suits and number of high cards (via a point value) – the more of the latter, the higher the hand value. Once a suit reaches more than '6 long' with the presence of A, K, Q and J, the hand becomes stronger. Probability calculations for initial hands are as above and an example for suit cards was given in Section 21.4.1. Extending these, we can answer one often-quoted probability question for bridge hands: '*what is the probability of getting a full suit set in a 13 card hand?*' For one particular suit, say all 13 diamonds, the answer has an extremely rare probability:

$$P = \frac{\binom{13}{13} * \binom{13}{0} * \binom{13}{0} * \binom{13}{0}}{\binom{52}{13}}$$

$$= 0.00000000000157477$$

'Any suit' (* 4) does not improve this much, P = 0.00000000000629908. Despite these low values, there are still reported incidences of the latter event in card games (www.dailymail.co.uk).

It is not until 'a mixture of suits' is indicated in events that we get to states that are more probable. These are obtained by slotting in the numbers in the above formula and ensuring that they total 13, e.g. a '4-4-3-2 *mix of diamonds, spades, clubs and hearts'*, respectively:

$$P = {}^{13}C_4 * {}^{13}C_4 * {}^{13}C_3 * {}^{13}C_2 / {}^{52}C_{13} = 0.0179593 \ (2\%)$$

This is a mix of named suits. Any mix of the four suits with this balance would be as 4!/2! = 12, which is the count for the number of ways that 4 suits can be slotted into 4 categories with 2 the same and 2 distinct: i.e. 4d,4s,3c,2h /4d,4s,3h,2c/4d,4h,3c,2s/ 4d,4h,3s,2c/4d,4h,3c,2s .....etc. In Excel, these calculations require function use as,

P('4-4-3-2')

= COMBIN(13,4) * COMBIN(13,4) * COMBIN(13,3) * COMBIN(13,2)

/COMBIN(52,13) * FACT(4)/FACT(2)

= 0.215512 (21.6%)

but the 'short hand' ${}^{n}C_r$ symbols are convenient.

The above ('4-4-3-2'), happens to be the most probable distribution of the suits in a bridge hand and all other distributions are of lower probability.

Surprisingly perhaps, '4-3-3-3', a more 'balanced' looking distribution is half the probability = 10.5%. As the number of a majority suit increases above 4, probability falls, ultimately to the very low probability above for 13 cards all in one suit.

Also useful is the 'long suit' - a majority of one suit, which along with high cards, affects bidding, at least for beginners. The probability for the dominance in the distribution is obtained by adding up all distributions of a particular dominance e.g. all hands with 4 as the maximum suit count are added. The '4 suit' is the simplest and there are only 3 cases with 4 dominant = 4 4 3 2 / 4 3 3 3/ 4 4 4 1 – adding these three gives 35%. The other cases have 6 or more combinations. Generally, '7 in the long suit' is good, but this and higher long suits are less probable – a long suit of 7 is approx. 4%, cf. 35% for 4 and 5 is the most probable at 44%.

Probability for high and low cards in a hand on the deal can be calculated. The most desirable high cards are the honours 'A', 'K', 'Q', 'J' +'10', with points allocated except for the latter. One of each in your initial hand, i.e. 'A,K,Q and J', (points 4+3 +2 +1 =10) has probability:

$$P = (^4C_1)^4 * {}^{36}C_9 / {}^{52}C_{13} = 0.037953 \ (4\%)$$

Other 10 points hands occur in other combinations such as 'A,A,J,J', 'K,K,K,J' and 'K,Q,Q,Q,J', etc. One way to get a count of the possibilities is by dipping into dice probability - the sequences for sums that total 10 of 3d4, 4d4, 5d6, 6d4 and 7d4 (Section 19.3.1) give the combinations, e.g. a 3-card sum of 10 honour points is given by the sequences of 3d4: a sum of 10 has 6 sequences 2 4 4, 3 3 4, 3 4 3, 4 2 4, 4 3 3, 4 4 2, but as combinations only 2 are required: 2 4 4 and 3 3 4.

Unfortunately, many more are found as the number of dice increase but the number of combinations doesn't increase in the same manner (5 in 4d4, 4 in 5 and 6d4 and 1 in 7d4). So, 16 combinations in total and the number of cards involved ranges from 3 to 7 (this latter combination consists of 'J,J,J,J and

Q,Q,Q'). These need to calculated individually and summed,

$$P = 0.0940511 \ (9\%)$$

This has the highest 'pointage' probability for an honours hand and it drops below 1% for 20 points. Similar calculations are used for other low or high hands, e.g. a hand low in high cards (a 'yarborough hand' named after the instigator the *Earl of Yarborough*, 1809 -1887)) has none above 10,

$$P = \text{COMBIN}(32,13)/\text{COMBIN}(52,13) = 0.000547033$$

Therefore, it's much more probable that you will have some high cards.

Blackjack

Another very popular casino card game, **Blackjack** (*'21'*, *Pontoon*) has a simple structure in terms of play. Each player receives two cards from the dealer (banker). Players attempt to beat the dealer's hand by getting as close as possible to a total of '21' on two or more cards. All picture cards are 10 and aces 1 or 11. 'blackjack' itself, an ace with a 10 or a picture card, is a hand that can only be equalled by the dealer getting one as well. In its simplest form, the game can be played recreationally at home with one deck of cards and these are not shuffled during play until a blackjack appears at which point the dealer can change.

Calculating initial hand probability for the first player concerns 2 cards. For a *blackjack* itself an ace (4 cards) plus a card with a value of 10 (16 cards) must be dealt,

$$P(\textit{'blackjack on 2, fresh deal'}) = {}^{4}C_1 * {}^{16}C_1 / {}^{52}C_2 = {}^{64}/_{1326} = 0.0482655 \ (5\%)$$

If the card value sum is below 21, further cards can be requested. A 'twenty-one' hand can appear with more than 2 cards, such as *'three 7s'*, although not often for the latter,

$$P = {}^{4}C_2 / {}^{52}C_2 * {}^{2}C_1 / {}^{49}C_1 = 0.000184689$$

The player must have two '7s' to begin with. All cards other than those of the player in question plus the dealer's up-card are treated as unseen and the second event is viewed as a selection of 1 from 49.

This player hand has a bonus payment in *Pontoon*, as do some other '21 or less' holdings, such as '6-7-8' and 5-(or more)-card cases (Scarne).

When we consider totals below '21' probabilities rise, e.g. 'getting 20' is achieved by 'ace + 9', or '10 +10',

$$P(\text{'getting 20 on 2'}) = (^4C_1)^2 / {}^{52}C_2 \text{ OR } {}^{16}C_2 / {}^{52}C_2 = {}^4/_{39} = 0.102564 \ (10\%)$$

A sum of '17', viewed as at or near a threshold in the game, can be attained by more combinations:

'10+7' (16*4 = 64 ways), '9+8' and 'A+6' (both 16 ways), which gives 96 in total and,

$$P(\text{'getting 17 on 2'}) = 96 \, /{}^{52}C_2 = 0.0723981 \ (7\%)$$
$$= {}^{16}C_1 * {}^4C_1 / {}^{52}C_2 \text{ OR } {}^4C_1 * {}^4C_1 /{}^{52}C_2 \text{ OR } {}^4C_1 * {}^4C_1 / {}^{52}C_2 = 0.0723981$$

It's possible to work out all the possible probabilities for any card combination in addition to these. Of particular interest are the ones at 'decision points' in the game.

Both you and the dealer can get a *blackjack* at the start, although it's much less probable. It can be calculated in various ways,

$$P(\text{'A blackjack to player then a blackjack to dealer'}),$$
$$= (^4C_1 * {}^{16}C_1 / {}^{52}C_2) * (^3C_1 * {}^{15}C_1 /{}^{50}C_2) = 0.00177302 \ (0.2\%)$$

Alternatively we can view the event as 4 cards in order, ' two As and two Ts', but the arrangements are limited as we must always have 'A,10' combinations, (thus 2*2 = 4 arrangements) and,

$$P = 4 * {}^4/_{52} * {}^{16}/_{51} * {}^3/_{50} * {}^{15}/_{49} = 0.00177302$$

Potential of the initial hand

For most card games, the *potential of the initial hand* is decided by its rating on the scale of pattern received or the content in relation to completion. With *Poker* games, there is an established ranked scale - the higher or nearer you are to a pattern on this scale the greater your chance of winning. Bridge has a point system for hand assessment, rummy games can be judge on the meld content or closeness to proposed melds. Blackjack hands are relatively simple to judge - if at a *blackjack* or close to '21' your success is almost certain or excellent, respectively, but other factors play a part.

Having looked at some of the 'on the deal' probabilities for the games above, a player can immediately make some estimates on the possibilities for other players.

Opponents' hand on the deal

Probabilities for the value of hands held by other players are based on the remaining cards excluding any that are seen and calculations tend to be more complex. Generally, all probabilities for players' hands in any card game are the same when viewed from the individual player perspective, i.e. based on a 52 card deal state. When judging from your own viewpoint, *the reduced deck (the deck minus seen cards) process* is followed (see below) with appropriate adjustments in some calculations. One way round this difficulty, is to simulate the deals, using the methods described in Book One Section 14.3, along with analysis that is more intricate. This can be set up for any card game and a fixed hand (yours) can be compared with random combinations for opponent(s) with analysis for content and counting of the incidence of 'better' hands.

Poker games

In *Draw Poker* games, after the deal only your hand is known. The higher the hand you hold on the scale outlined above, the more chance you have of success and <u>generally</u> the lower the probability that an opponent(s) will have a higher hand (it depends on the number of players - more players mean more chances for a hand to exceed your one). For example, on the deal you have one of the commonest hands – a pair of '9s' – what is the probability that your opponent(s) has a better hand? For this, all forms of hand that exceed a pair of '9s' must be included, i.e. not only two pair, 3 of a kind, flushes, etc. but higher pairs and even the other pair of '9s' combined with a higher 'kicker'. To get the wider probability of 'any better hand' would require probabilities for all higher hands from 3oak upwards. Calculation for this particular type of problem can get quite involved.

An easier example would be if you were lucky enough to be dealt 'four of a kind' in 2-player game. A rare hand on the deal, but it's even rarer for two players to be dealt such a hand as the following workings show. Assume your hand is 'four Ks and a $Q\clubsuit$'. This is a 'mixed subgroup' event and the probability calculation is based on (21.2c),

$$P('K,K,K,K,Q\clubsuit') = {}^{4}C_4 * {}^{1}C_1 / {}^{52}C_5 = 0.000000384769$$

(As a check, note that $P('K,K,K,K,Q')=0.00000153908$ and dividing this latter value by 4 gives the above)

To have a higher 4oak than this, your opponent would need 'A,A,A,A,x' and,

$$P('get\ 4\ aces\ and\ any\ other\ card') = {}^{4}C_4 * {}^{43}C_1 / {}^{47}C_5 = 0.0000280324$$

The opponent gets the hand from 47 cards and '*four Ks with a $Q\clubsuit$*' cannot be beaten by any other 4 of a kind except 4 aces. There is but one way to get the As and the 'kicker' can be any of the remaining 43 cards. So, the probability of a higher 4oak is very low (0.003%) and other hands that would beat your 4 Ks include straight flushes, which have even lower probabilities.

These calculations can be extended for more opponents, typically resulting in an increased chance that your hand will be exceeded (see other example below). Bărboianu provides details and extensive tables on these probabilities for draw poker and other forms and poker odds calculators are available which will give probability for one or more opponents having a higher hand than yours (probability.infarom.ro). Generally, certain hands win over a higher percentage of the time than others – but these relate to the occurrence of these hands in deals - viz. one pair, which occurs more often, is likely to win more games than any other hand.

Similar procedures can be used to estimate opponent hands in other forms of poker but the number of visible and face-down cards varies. Thus, in *Stud Poker* only one hole card and one up card appear early in the game, in *Texas Hold 'em* there are two down cards. These games have more betting rounds making valid judgements to 'stay in' more crucial.

The latter game has some relatively simple calculations for opponents' hand strength, e.g. if you have a pocket pair of '10s', the probability of one opponent having a higher pair ('J, Q, K or Ace') before the flop, is,

$$P = 4 * {}^{4}C_2/{}^{50}C_2 = {}^{24}/_{1225} = 0.0195928 \ (2\%)$$

There are ${}^{52}C_2 \ (=1326)$ deals of 2-card hands but for the opponent, the deal is based on ${}^{50}C_2 \ (=1225)$. There are four possible ranks that exceed '10' and we are taking two from each. The resulting probability is low (2%) and you might feel confident, but as more opponents are brought into the calculation the chance of a higher hand increases. With two opponents, the reasoning is that the other players can beat your hand as:

'p2 has higher' OR 'p3 has higher' OR 'BOTH have higher'

An approximate probability can be obtained by applying the ***addition rule*** for non-mutually exclusive events (Book One Section 5.2.2), with formula (5.2b),

$$P = 0.0196 + 0.0196 - (0.0196 * 0.0196) = 0.0387998 \ (4\%)$$

As more opponents are included, probability will be given by,

P = no. opp.* 0.0196 - all the joint probabilities

Admittedly, the joint probabilities (2-way, 3-way, ...) are subtracted, but these get smaller and smaller as the no. of opponents rises, so the largest contribution is the base probability. For seven opponents, the approximate value is,

P('*at least 1 opp. has higher hand*') = 7 * 0.0196 = 0.137143 (14%)

Although the true value is slightly lower (13%), this is a considerable increase in danger from 2% against a single opponent. Now, you do not feel quite so confident with your pair of '10s'.

The potential of initial pre-flop hands in *Texas Hold 'em* can be obtained by very complex calculations (en.wikipedia.org) or more conveniently by simulation. Using the latter technique as described above, a fixed initial pre-flop hand (player 1) + all the random community cards is compared with the random 2 cards of player 2 and the shared cards, for the best 5-card poker hand. One thousand deals were simulated and for a pair '10s', approx. 75% resulted in player 1 having a better hand than their opponent did (random hand).

Doing this for a variety of pocket hands shows that '2 aces', at 85%, have the greatest potential, followed by combinations containing one or two 'high' cards. Generally, potential decreases with rank, suited and close linked nature, bringing the percentage down below 50%. Least potential is found with low cards and off-suited (33 to 39% cf. 22 to 49%), exact figures depending on 'off suit' or 'suited', etc. Usually, the presence of one high card, '9 or better', keeps the probability of winning above 50%, but then it falls below. Values are for one opponent and decrease with more. Consequently, the pocket pair of '10s', although good to begin with, may have to be looked at in a poorer light if there are several players. More sophisticated calculators are available on the web handling more opponents (holdemtight.com).

An enhancement of the simulator quoted above, allowed comparison of two fixed pre-flop hands (player A and B) in terms of potential and an illustration of *hand dominance* (wizardofodds.com). This feature of the game is where opposing hands are similar but one dominates the other in that the dominated hole card hand has limited ways to beat the other. A clear example is where a pair is always dominated by a higher pair even if just 1 rank lower: 'A,A' vs. 'K,K' - the aces dominate and win 81% of time, the kings only 19% of time, but there is little difference for all lower pairs: 'two 2s' wins 18% of time. Then, pairs always dominate non-pairs where there is a match or higher rank with one card of the non-pair, etc.

Another aspect of dominance is where an apparently weaker hand performs better than a stronger one against an opponent's higher hand. Consider the situation where one player (B) has hole cards 'K♣,A♠' and player A has a very similar hand of 'Q♦,A♥'. On the surface, the latter hand looks a reasonable competitor. Running a comparison over 1000 deals showed that player A's 'Q+A' has a 25% estimate of a win and in this case B's hand dominates A's as there is a match with one of B's cards. Any cards on the board (community) that help A, also help B (excepting 'Qs'). Removing this dominance has an immediate improvement of A's chances. If A had a hand close (in terms of ranks), but with no matches, such as 'Q♦,J♥, it would have an increased a chance (33% estimate) as the hand is no longer dominated. Even very low off-suited cards perform better than the much higher, but dominated 'Q♦,A♥'. Thus, while the hand potential described above does not change because of an opponent's hand, some hands can do better than others can.

Of course, any individual player cannot see the opponent's hole cards but may glean information via pre-flop betting, and play is enhanced by being aware of hands that can be dominated.

Rummy

With *Rummy* games, simple probability reasoning based on the balance of various subgroups of the cards can tell you about what your opponent(s) has, especially with large hand games such as *Gin*. Thus, if you have a lot of 'red' then your opponent has a higher chance of having more ' black' than 'red' and similarly for 'odd/ even 'and 'high/ low', etc.

For example, at the start of a game of *Gin rummy*, you have 8 red in your hand, does your opponent have more? There are 18 red cards available (ignore discard but assume not red) and we can work out the probability of having 9 or more as,

P('*opp. has more red than you'*) = 1 - HYPGEOM.DIST(9-1,10,18,42,1)

= 0.000822765 (0.1%)

A rare chance and an exact match is also low in probability (P('*10 red each'*) = 1%)). It is almost certain that your opponent has less red,

P('*opp. has less red than you'*) = HYPGEOM.DIST(7,10,18,42,1)

= 0.990970 (99%)

A few deals of the hands will usually show this - deal yourself one hand with 8 red in 10 cards then deal out the opponents hand and repeat the process 10 times (one run of this gave from 0 to 6 red cards on 10 deals). We can calculate probabilities for each count and graph them. The first hand is based on 52 cards (all red present), the second on 42 cards with 18 red present(Fig. 22.1).

Therefore, the red content in the first hand has a mode of 5, but for the second hand the mode is lower at 4. The distribution of the no. of reds in hand for the remaining deck low in red is 'skewed' (off-centre), it has a flatter R-tail and falls off more quickly signifying a relatively lower probability for more reds.

Thus, player 1 knowing these probabilities can assume a lower count of reds and corresponding more blacks. The opponent will be more likely to require black ranks for runs, etc. and player 1 should try to limit access to these in discards.

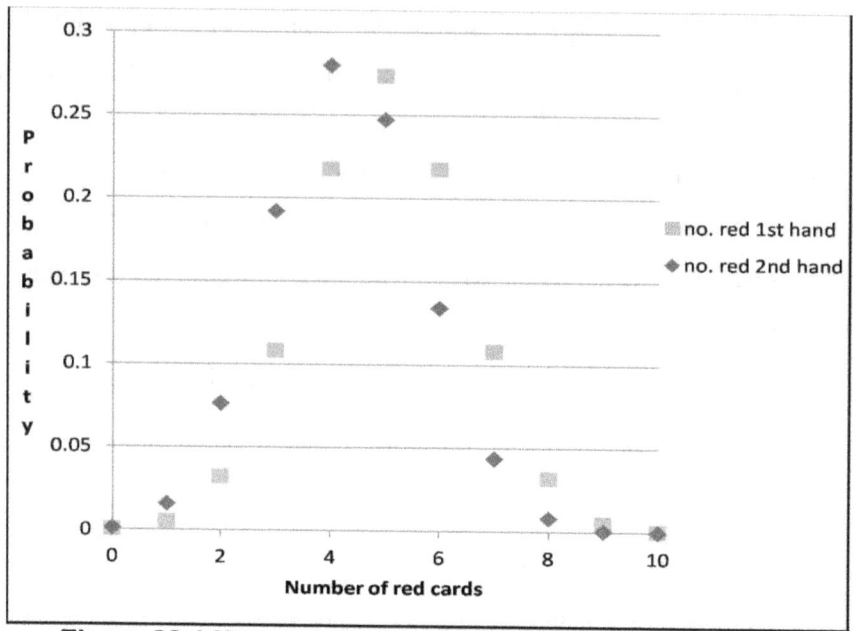

Figure 22.1 Hypergeometric distribution of the number red cards in 10 card gin rummy hands

Whist and bridge - distribution within hands

An approximate idea of the composition of opponents' hands in *Whist* and *Bridge* before play can be gauged by referencing the most probable patterns on the deal, some of which are calculated above. Knowledge of the probabilities of these initial hands allows players to judge opponents' hands. The more balanced distributions are more likely, dependent on known suits within individual opponents and between them. (www.bridgehands.com).

As all cards are dealt out in *Bridge* and 13-card whist versions, we can calculate the probability of any particular division. Referring to the partitioning of objects (Book One Section 7.4), we have the number of ways that sets of 13 cards can be dealt out to 4 players from 52, namely, $52!/(13!13!13!13!)$. We now need the number of ways for particular divisions and we can use the same methods, e.g. take the split of a particular suit (spades) across the 4 hands as 7 2 3 1,

$$P = 13!/(2!7!1!3!) * 39!/(11!6!12!10!) / 52!/(13!13!13!13!)$$

$$= 0.000783679 \ (0.1\%)$$

The other cards in each hand are a partitioned selection from any of the remaining 39. A more balanced spread of the spades such as '3 4 3 3' is more probable (0.026; 3%).

Blackjack

In *Blackjack* the opponent is the dealer and the main interest after the deal is judging the dealer's hole card and his probability of getting close to '21', *blackjack* or busting. You know your 2 cards and the up-card of the dealer – but what's the hole card? Take the case where you have 'Q,3' and the dealer has a '7' showing – what is the probability he has ace based on a reduced deck size of 49 cards?

$$P(\textit{'dealer has an ace'}) = {}^4C_1 \, / {}^{49}C_1 = {}^4/_{49} = 0.0816327 \ (8\%)$$

This applies to any particular rank other than the 'Q's or '3s'. Extending this to get the probability that will put the dealer into a strong position (typically this will be above '15'), we can count up the cards that give a total of '15 -21',

No. = A, 9, 10, J, Q, K = 6 * 4 -1= 23

$$P = {}^{23}/_{49} = 0.469388 \ (47\%)$$

(The closest the dealer can get is '18' - a '7' plus an ace)

So, the hole card could be one of 23 cards and any one of these means that the dealer will stand and you would need to get '19 or more'. Of course, it all depends not just on the up-card, but on your cards too. Some prepared tables standardise this by giving probability based on 51 cards (www.blackjackinfo.com).

22.2.2 Probability during exchange of cards and play

Once the card player has evaluated his/her initial hand and made some estimates on those of the opponents', they can look to the play. In many card games, a stage of *exchange* occurs where players discard one or more of the cards in their hand and replace them with fresh cards from the deck stock or sometimes from a face up discard pile.

As pointed out above, other aspects of the play can reveal crucial information not least any further cards that are revealed. In addition, your position at the game table and activities of other players, all have a bearing. Significant clues can appear here and the psychology of 'reading' the other players actions or lack of, are crucial to these aspects of the game. Returning to the exchange process, choice and probabilities enter the decision making in the game.

This is typified by the continual choices during *rummy* play as new cards appear on the discard pile. In other games, as with the draw in *5-card* and *Video Poker,* the sequential revelations in *stud* and *Texas hold 'em* ('flop-turn-river') and *Blackjack* turn over, all provide information for re-assessment of probabilities. In all circumstances of this type, the following points apply:

- It does not matter how many other players there are as all unseen cards, including those of these players, are viewed as the 'remaining cards'

- The draw comes from the remaining cards, although physically it comes from the reduced pack

- Subsequent probability calculations must take into account all cards known to the player (i.e. his/her hand, other player revealed cards, any face-up game cards and discards)

So, in poker games opponents' hands are judged on 52 minus the no. seen = 50 or less cards, in 7 card rummy or whist on 52-7 (8 if include first up-card), etc. Thus, for *5-card draw poker* the remaining cards will always number 47.

If a choice is made to exchange 3 cards in the draw then this is viewed as a selection of '3 from 47'. For 3-card games, it would be 49, in two-player *canasta* (11-card) with 2 decks plus jokers it would be 86, etc. (Table 22.3). Selections are therefore from a reduced deck and enumeration of outcomes must consider the cards kept and those discarded in the calculations. Many of these points are discussed in sources on card game strategies although probability may not be mentioned as such.

Rummy

Simple exchange probabilities are found with **rummy games** where continual exchange occurs within the play and substitution usually comprises single cards. Known cards in the hand depend on the game version, straight rummy having fewer (5-7 cards) than *gin rummy* (10). Of, course, if the rules allow melding before the end then players will have smaller hands, e.g., 4 cards remaining to form the final meld. Larger hand games provide you with more information about your opponent's hand. If you have more of a particular suit, rank or subgroup, as shown with the red cards above, then the probability for your opponent to having the balance of the remainder would be or interest. For example, in a *Gin* deal, you have 'two Ks, two 5s and two J' and there is another 'J' as the first discard. The probability that your opponent has the other members of the ranks is relatively simple if we assume combination with any other cards,

$$P(\textit{'two Ks, two 5s, one J'}) = (^2C_2)^2 * {}^1C_1 * {}^{36}C_5 / {}^{41}C_{10} = 0.000336270$$

Very low, but for 1 of each the figure is higher,

$$P(\textit{'one K, one 5, one J'}) = (^2C_1)^2 * {}^1C_1 * {}^{38}C_7 / {}^{41}C_{10} = 0.0450281 \ (5\%)$$

Table 22.3 Improvement probability in some card games

Initial hand	No.[1]	Target	Calculation	P
Rummy 7-card: J,J,-,-,-,-,-	1 from stock	JJJ set	$^2C_1 / {}^{36}C_1$ (with 9 card discard)	0.0555556 (6%)
Gin hand: K,K,6,6,T,T,3♦,4♦,-,-	1 from stock	Run ♦s open OR set 10s, 6s OR Ks	$(^2C_1 * 3 + {}^2C_1) / {}^{28}C_1$ (with 14 card discard)	0.285714 (29%)
Canasta 2-hand (11): **(jks,2s wild (108)** (5 card in hand + meld of 8,8,jk,8,8,8 laid down) *jk = joker*	1 from stock	Mixed Canasta of 8s with an 8(3) or 2(7) or jk(3)	$(^3C_1 + {}^7C_1 + {}^3C_1) / {}^{(108-30)}C_1$ (with opp. lay own of five, Ks,2; 13 discards)	0.166667 (17%)
Draw Poker: One Pair	3	Two pair	[a] see below	0.159852 (16%)
			$^a ({}^9C_1 * {}^4C_2 * {}^8C_1 * {}^4C_1 \text{ OR } {}^9C_1 * {}^4C_2 * {}^3C_1 * {}^3C_1 \text{ OR } {}^3C_1 * {}^3C_2 * {}^9C_1 * {}^4C_1 \text{ OR } {}^3C_1 * {}^3C_2 * {}^2C_1 * {}^3C_1) / {}^{47}C_3$		
One Pair	3	3oak	$(^2C_1 * 927) / ({}^{47}C_3)$	0.114339 (11%)
One Pair	3	full house	[b] see below	0.0101758 (1%)
			$^b({}^9C_1 * {}^4C_3 \text{ OR } {}^3C_1 * {}^3C_3 \text{ OR } {}^3C_1 * {}^3C_2 * {}^2C_1 \text{ OR } {}^9C_1 * {}^4C_2 * {}^2C_1) / {}^{47}C_3$		
4 open str., 2 ends	1	5 straight	$2 * {}^4C_1 / {}^{47}C_1$	0.170213 (17%)
Video poker: Treble	2	4oak	$^1C_1 * {}^{12}C_1 * {}^4C_1 / {}^{47}C_2$	0.0425532 (4%)
4 flush	1	5 flush	$^9C_1 / {}^{47}C_1$	0.191489 (19%)
Texas Hold' em: Pocket pair 8s	3 (flop)	Set 3oak	$^2C_1 * {}^{48}C_2 / {}^{50}C_3$	0.115102 (12%)
Stud Poker : **7-card: 2-hand** (one K hole card, no other Ks seen)	6 per player	Another K appears in my hand	$^3C_1 * {}^{48}C_5 / {}^{51}C_6$ (treats other player cards as unseen)	0.285234 (29%)
Blackjack: **'16' (J,6)** (11 cards out (inc. above, A,4,4), dealer up card = 8)	1	21 or <	three As, four 2s, four 3s, two 4s, four 5s $(3+4+4+2+4)/(52-11)$	0.414634 (41%)

[1] number of cards changed or added

As seen above, complete melds have relatively low probability on the deal but 2-card incomplete versions, as in the latter example, are more common.

Choices come down to selection of an unknown card from the top of the face up discard pile or from the top of the stock. Usually, unless the discard will complete a meld, drawing the unknown stock card is more advisable. This latter draw has a probability relative to the unknown cards and the possible melds in your hand cards. Calculations are relatively simple cf. the *Poker* conversions below, and the chance of converting various rummy melds are,

$$P = {}^rC_1 / {}^nC_1 = r/n$$

where n = all unknown cards (remaining deck plus other players held cards)

r = remaining unknowns of interest

If one particular card is required, as in converting 2 of a kind to 3 of a kind, where a player has 5 cards ('Q,Q,5,9,K'), there are 2 other players, each with 5 cards, 3 cards face up (discard pile) and 34 in the stock pile. Probabilities are based on selection from all unknown cards, viz. the other players' hands and the stock = 44. Thus,

'getting a Q on the next draw from the stock, (with no 'Qs' in discards)' –

$$P = {}^2C_1 / {}^{44}C_1 = 0.0454545 \ (5\%).$$

With 1 'Q' in discards this is reduced to 2%.

Relatively low probabilities, but remember if you have another '2oak' this would double your chances for the same circumstances, even with a 7-card hand you could have three 2-card sets. As more cards are exposed, probability increases - say that now (example above) there are 17 in the stock, 20 in the discards but you have not seen a 'Q' being picked up by an opponent, then,

$$P = {}^2C_1 / {}^{27}C_1 = 0.0740748 \ (7\%)$$

An improved possibility to pick up the 'Q'.

Runs (straight flushes) can have more possibilities for completion than sets, as long as the straight has open ends. Now, two cards qualify and this is maintained with the open straight. We can see that this potential is more so for cards of medium rank ('6 to 8s'). With the play example above similar data can be used to illustrate - an 'inside run' - '6♦,8♦', has but 1 card (P = 2%), whereas 'open-ended', '7♦,8♦', has two (5%). With more cards in the hand, as in *Gin rummy* and *Canasta*, you could have several pairs or partial runs. Thus, increased incidence of unfinished melds means that the probability of achieving a useable set or run increases. The picture is complicated as some may overlap - however you are limited by the fact that in basic game versions you can only chose 1 card.

Part of the strategy in such games is to recall discards, left and collected, as a decision guide to your own collection path. Judgements on 'which card to discard' depend on what you have and what you have seen and what your opponent picks up and discards. This knowledge will affect the probability of your own intentions and you will be able to see how you may be able to lower the probability for your opponent.

One important step is where you have a card that is of no use to you, but can your opponent use it? This choice can be weighed in terms of a dichotomous probability, i.e. the opponent 'needs it or not'. The positive prediction is based on previous discards and action by this player - if seen to have drawn such a card or related indicates 'yes' - if she/he has discarded or ignored on the discards then 'no'. In addition, the balance of card groups must be taken into account. Unfortunately, you may not have other options for a discard. As for your own actions, if you have seen a 'K' go into the discard pile and you draw a 'K', then you know that your chances of using that 'K' in a 3oak are less than if you had not seen it. Keeping that 'K' and hoping to pick up the other two,

$$P(\textit{'the other K'}) = {}^2C_1 \, / \, {}^{41}C_{10} = 0.048780488$$

This does not complete the set and you would need to have a follow up after 2 other players go round - say 2 more on discard, then.

$$P(\textit{'final K'}) = {}^1C_1 \, / \, {}^{39}C_{10} = 0.025641026$$

The probability for both these events happening is their product, P =

0.00125078.

If no 'Ks' were seen in the discards, the probabilities are 0.0731707 (one 'K'), 0.0512821 (2nd 'K') and the joint is P = 0.00375235. Still very low but four times better than hoping to get '2 from 2'.

In game versions where melds are laid down during play, some judgements can be easier. How many cards do they have remaining in their hand? If they have only 1 card then they cannot make a meld in-hand. For example, you have a '5' – there are no '5s' in the discards and no possibility of the card being used on the laid out melds - so you are safe to discard the '5'.

Whist games

Play in *whist* games relies on some probability- aided decisions, plus the information revealed by opponents' bidding and subsequent play. For *Bridge* in particular, there is an enormous bibliography and most include probability. A simple account is presented here, via judging opponents' hands in more detail, via bidding and exposure of the dummy hand and some consequences.

After the deal, you can judge what other opponents have initially by a count of what you have in the same way used for rummy games above. All cards are deal out, so if you have nine spades in a whist game, then there are only 4 elsewhere. However, referring back to the calculations above (Section 22.2.1), 'nine of a suit' is much less likely than is 'four of a suit'. In bridge, with you and your partner, a nine count is much more common.

Looking at bidding even in a simplistic way reveals more - the higher the bid the more of that suit along with more high cards. There are also conventions in bidding that allow 'signals' between partners, although these are not secret, to allow formulation of the best bid. Based on your own hand and that of your partner can allow some estimates of the opponents' holdings. Consequently, if you have a long suit of 4 of a suit and your partner has 3 of the same then there are only 6 left, distributed between your 2 opponents. If you examine a full list of all the distributions (www.bridgehands.com), you will notice that some distributional splits are more likely than others are. It can arise that a pair of players (N & S) know, based on their own hands, that their opponents have 5 of a particular suit - how are these cards divided up into 2 sets for these other players? The partitioning can be as containing from 5 to 0 in a set, i.e. make 1 choice and the other is automatic as 5,0 / 4,1/ 3,2 / 2,3,1,4 / 0,5. Taking the cards as distinguishable and looking at them in isolation, they can be split in various ways:

'5,0' (or '0,5') split, $^5C_0 * 2 = 2$ ways

'4,1' (or '1,4') split, $^5C_4 * 2 = 10$

'3,2' (or '2,3') split, $^5C_3 * 2 = 20$ Total $= 32$

In this example, all of the possible splits can occur in two permutations, corresponding to W and E opponents (unlike with 6 cards where there would be a single '3,3' split). The probability of a random allocation of 5 cards into two groups giving a split of '4,'1 or '1,4' is $^{10}/_{32} = 0.312500 = 32\%$ and 6.25% and 62.5% for the '5+0' and '3+2' divisions, respectively. Therefore, this latter breakage is massively more likely than the others are.

Calculating these divisions with bridge hands rather than the isolated cards can be done by using the partition calculation as above, and,

P('3,2 or 2,3 split of suit in 13 card hands')

$= 2 * 5!/(3!*2!) * 21!/(11!*10!) / 26!/(13!*13!) = 2 * {}^5C_3 * {}^{21}C_{10} / {}^{26}C_{13}$

$= 0.678261 \ (68\%)$

Similar calculations are followed for any split based on the knowledge of 'your side' in the game. We know this latter count with 100% probability, in the example, that W and E have 5 of the suit in question. The partition calculations tell us the most likely split(s), i.e. the ones that cover most of the sample space. This information can be used to guide play as it is, but more important is the location of any honour cards in opponents' hands. Additional clues can be gleaned possibly from bidding and later play, but the initial leads are critical.

Once bidding is complete, fresh light can be shone on the breakage of the outstanding cards. Suggestion of a high count of a second suit by one of your opponents could mean that have less room for other cards and the split is less likely to be the most probable random one. For example, E has bid for a 4 spade contract, suggesting a hand high in spades making splits with a higher content in E less probable, more likely to be' 3,2' or even '4,1'. Initial play could reveal more.

Bridge techniques such as the *finesse* make use of this information, to overcome possible opponent advantage due to a gap in your partnership's high card holding. Probabilities for success depend on the number of cards in suit held by your side and how the outstanding cards are split. More detail is given in reference sources (www.durangobill.com) for different numbers of suits. The more even splits have generally higher chance of success than the uneven splits and where there are fewer cards possessed by the opponent that you lead to.

Poker exchanges and play

Contrasting with single card exchanges, in *Draw Poker* up to 5 cards can be exchanged, based on the initial hand and the particular target pattern(s). Probabilities for making the target(s) can be calculated to aid decisions. If changing one card then calculations are similar to some above, but again the draw may have potential for more than one target. Thus, drawing one card to a flush or straight could achieve these targets but could also result in a pair. If you possess a pair within the hand to begin with, there are more possibilities.

For an illustration of this, imagine a player, Pedro the poker aspirant, in a game of draw poker with '4♣,4♥,J♥,A♥,10♥' on the deal. There are 3 other players each with a hand of 5 cards. It's useful to classify the potential of the dealt hand to provide a basis for decisions - this is a **'4-flush, single pair, 3-inside straight'**. How can we advise Pedro? Should he keep the hand and stand pat? Or, should he change 1 or more cards? He could keep the two fours and discard the others. Or he could throw out just the 'J,10' and keep the '4s' with the ace? There is a range of possibilities - which one has the highest probability and potential?

Before answering, take on board the above points - the probability is based on 47 cards (5-card draw), minus any that you see on the exchange. The calculations are slightly different because of this. Essentially, we are back to a situation of selection of named cards with unspecified cards (Section 21.5), with different parameters - the deck size is 47 and from 1 to 5 cards can be drawn.

Each particular option from the available, results in one or more hand outcomes, some or all of which give an improved hand. Returning to our impatient player, before advising him we need to first identity these options and calculate the probabilities. The most obvious one is to retain the pair and change 3 cards, and for this first improvement option more detail is given.

<u>Option 1</u>: Keep '4♣,4♥', discard the other cards and draw 3

Assessment of this option will give the probability of improving his 'single pair' hand. Possibilities are to get another '4' or 'two 4s' out of the pack, up to another pair or 3oak, etc.

The simplest case in terms of calculation is found the probability of getting another two '4s', i.e. four 4s in total (a '4oak' hand). The target combination is '4,4,x' where x can be any other card. We have no problems with repeated ranks cropping up as there is a single unnamed card, and we need not consider the discards as, in this calculation, they cannot influence the choice:

Target 1: '4,4,x' to get '4oak'

$$P('two\ 4s\ in\ a\ selection\ of\ 3,\ '4,4,x'\ from\ 47')$$
$$= (^{2}C_{2} * {}^{45}C_{1}) / (^{47}C_{3}) = 0.00277521\ (0.28\%)$$

This is explained as the choice of 'two 4s' from the 2 which remain somewhere in the 47 cards, and the choice of the unnamed card comes from the remaining 45, after the '4s' have been selected. There is a very low probability of picking up those extra two fours and the player would be advised not to pin his hopes on such an event. However, this is not the only possibility.

'Getting one 4' sounds more probable, but the calculation is not quite so straightforward. The catch here is that the selection is '4,x,y', where $x \neq y \neq 4$ and repeated ranks are a possibility – we cannot simply calculate the choice of the 2 unnamed cards with $^{45}C_2$. Following the procedures used above (one pair calculation) we must partition the selection as selection of a single '4' ($^2C_1$), then select 2 unnamed cards from the remaining 45 cards (exclude the other '4') as subgroup ranks. There are 12 rank subgroups but in this case 3 of them are deficient due to discards, namely 'J,10A' – thus we have 9 'full rank' subgroups. This second step gets complicated as there are several ways in which these 2 cards can be selected, as cards can be drawn from the full rank groups or from the deficient ones, or 1 from each:

$$= {}^9C_2 * {}^4C_1 * {}^4C_1 = 36 * 4 * 4 = 576$$
OR,
$$= {}^3C_2 * {}^3C_1 * {}^3C_1 = 3 * 3 * 3 = 27$$
OR,
$$= {}^9C_1 * {}^4C_1 * {}^3C_1 * {}^3C_1 = 9 * 4 * 3 * 3 = 324$$

As the 'OR' operative is applied we can sum these to give 927.

Target 2: '4xy', $x \neq y \neq 4$ to get 3oak'

$$P('one\ 4\ in\ 3\ cards') = ({}^2C_1 * 927) / ({}^{47}C_3) = 0.114339\ (11\%)$$

So, although still not 'good', there is a much improved chance of getting 3oak than going for 4oak.

The process is not over yet and other hands are possible (see Table 22.3 for calculations):

Target 3: 'x,x,y', $x \neq y \neq 4$ to get 'two pair'

This calculation involves adding up various selections from full and partial ranks as described above,

$P = 0.159852$ (16%)

Target 4: 'x,x,x', $\neq$ 4 OR '4,x,x', x $\neq$ 4 to get 'full house'

$P = 0.0101758$ (1%)

The possibilities for each target are summed to give overall *improvement probability*,

P(option 1) = 0.28 + 11.41+ 16 + 1 = 28.5%

Pedro can also take another option: changing the '4♣' allows a go at a flush in hearts,

<u>Option 2</u>: keep '4♥, J♥,A♥,10♥' discard the other '4' and draw 1 card

The target here is the flush of course but there is another possibility in that he could fail and end up with nothing, or he has a chance of getting a pair the same as or higher than the original.

Target 1: 'x = ♥' to get flush, $P('flush') = {}^9C_1 / {}^{47}C_1 = 0.191489$

Target 2: 'x' = '4,J,A, or 10' to get pair ideally not = '4'

'*A higher pair than 4s'* ($\equiv$ '*get a J OR a A OR a 10*')

$P = 3*{}^3C_1 / {}^{47}C_1 = 0.191489$

Overall for option 2,

$P('flush \ OR \ higher \ pair') = 0.382979$ (38%)

[A 'Js or better' game would reduce this probability to $\approx$35%).

<u>Option 3</u>: Keep 'J♥,A♥,10♥' and draw 2 cards

Target 1 'x,y' = any 'Q,K' to get straight or 'Q♥,K♥' to get royal flush - the top hand, or get two '♥s' for the flush. Unfortunately, as we shall see below in the next section, attempting to convert a '4-inside straight' is 'difficult' enough, but here Pedro would be starting with a '3-inside straight',

$P('3 \ inside \ straight \ to \ 5 \ straight') = {}^4C_1 * {}^4C_1 /{}^{47}C_2 = 0.0148011$ (1%)

However, this includes a royal straight flush and to get the basic straight we require to exclude hearts and,

$P('Straight\ without\ K,Q\ ♥s') = {}^3C_1 * {}^3C_1 / {}^{47}C_2 = 0.00832562\ (0.8\ \%)$

$P('3\ inside\ str.\ to\ royal\ flush') = 2/{}^{47}C_2 = 0.000925069\ (0.1\%)$

These are mutually exclusive and to get the union, for both by addition,

$P = 0.00925069\ (0.9\%)$

This is a lower value than above because the former also counts a straight with 'K or Q ♥s' with the three other 'Ks and Qs'.

Similar complications arise with getting a flush - nine '♥s' in total are available, but we cannot have 'K♥' and 'Q♥' together, so in effect there are 8 and,

$P('3\ flush\ to\ 5\ flush\ with\ K♥') = {}^8C_2 / {}^{47}C_2 = 0.0259019\ (3\%)$

OR

$P('3\ flush\ to\ 5\ flush\ with\ Q♥') = {}^8C_2 / {}^{47}C_2 = 0.0259019\ (3\%)$

So, net probability = 0.0518039 (5%)

This option still has more and there is a chance of pair, two pair or 3oak, calculated simply as,

$P('pair') = (3 * ({}^3C_1 * {}^{44}C_1) + ({}^2C_2) + (9 * {}^4C_2)) / {}^{47}C_2 = 0.0758557\ (8\%)$

$P('two\ pair') = (3 * ({}^3C_1 * {}^3C_1)) / {}^{47}C_2 = 0.0249769\ (2\%)$

$P('3oak') = (3 * ({}^3C_2)) / {}^{47}C_2 = 0.00832562\ (1\%)$

So, P = (1 + 5+ 8 + 2 +1) = 17% overall for option 3 conversions.

Other options are a possibility - keeping 2 cards or 1, or changing all 5 for a replacement hand. The former options would give even lower probabilities but are sometimes used when a hand has 'nothing' expect for 1 or 2 high cards. Without even these, full replacement is the only possibility or a fold.

So, Pedro has around 29 + 38 + 17 = 83% chance of improvement, split

50:50 between the higher ranked flush and one pair (it could be a pair of 'As').

We can advise Pedro now. Based on probability alone we could say that Opt.1 =28.5 Opt. 2 = 38% and Opt. 3 = 17%. Option 2 has the most potential, but there are several other factors to take into account. One is the *risk level* in each option if it fails. For option 2 and 3 the risk is high (he gets nothing if conversion fails), whereas with option 1 Pedro will still have his pair (assuming '4s' are enough to stay in). The next consideration is how many players are in the game and what position is Pedro in the order of play? Are there any 'tells' from the other players on the first round of betting? On the draw, how many cards did the opponents change -1 card, 2 or 3 cards, etc.?

His own actions can give information to the other three players - changing 3 cards says 'I have a pair or I have nothing and I'm bluffing'. So, Pedro could keep the '4s' plus the ace kicker - the opposition will wonder - does he have a 3oak or is he trying for an outside 3 flush /straight conversion? These are just a sample of the many issues that are discussed in texts on strategy and tactics in poker play. Additionally, the probabilities themselves are not the whole story - expected values tell more (see example below in video poker).

These steps can be applied to other types of draw and exchange in poker and other card games (Table 22.3). Improvement and exchange probability exercises appear in the Problems and Exercises pages.

Texas Hold 'em

With some other poker games, the discard and improvement stages do not occur as such and some cards are dealt face up for all to view. Now probabilities can be judged on a lot more seen cards, e.g. in 5-card *Stud Poker* with four players, 16 cards can be identified if all participants stay in the game. In *Texas Hold 'em*, two cards are dealt out to each player and a common pool of seen cards is available for each player to make up the best hand. Probability problems are abundant, especially in this latter game and not surprisingly, there are many, many books written on the subject. Because of the sequential revelation of cards and the betting structure, play consists of a continual process of assessment for 'staying in the game', which gets more committing as the pot grows in size.

In *Texas Hold 'em*, once you have your initial two cards you can consider their potential prior to the flop stage, i.e. overall chances of winning and the probability of improvement. If there are no obvious possibilities, you have 'nothing' (such as low cards, different ranks, non-suited, non-consecutive) then the advice is usually to fold (dependent on the no. of players and your position). As seen above, the most common pocket hand (24%) is a 2-flush, i.e. two suited cards - so what's the potential for a full flush? There are 50 unknown cards and 5 to come to the board as the community cards - you need 3 other of the particular suit out of these to convert and,

P(*'3 or more of suit'*, *community cards*)
= 1-HYPGEOM.DIST(3-1,5,11,50,1) = 0.0639983 (6%)

This is low, but there are many more possible ways that you could end up with winning cards for your hand.

Other starting hands have much greater potential, in particular, 'pocket pairs'. They have a lower initial deal probability (6%) but convert to a treble set as,

P(*'particular pair to set'*) = HYPGEOM.DIST(1,5,2,50,0)
= 0.183673 (18%)

Moreover, this is not the final assessment as it has potential for two pair and a full house. Generally, pocket hands have one or more possibilities for improvement but overall a pocket pair has most potential.

There are some surprizes in this initial hand capacity due to the sharing of the community cards. This gives rise to cases where both players' hands benefit from the flop and later cards but one dominates the other as it has a higher rank in one card, e.g. 'A,Q' vs. 'A,K' (see above).

We have looked at the potential of the pocket cards. These probabilities change as the community cards are revealed. Once the flop is dealt, the player must be careful not to be drawn in by the appearance of the cards alone. You need to judge your hand's potential and estimate those of the opponents' and take notice of the betting. You can count the 'outs' for your hand and compare these to the pot odds (see below). With more opponents, there is higher probability for at least one higher hand.

Probabilities of improving your hand on the flop are calculated for a deal of 3 cards from 50, e.g. you have a pocket pair, then the probability of another pair on flop is,

$$P = COMBIN(4,2)*12 *COMBIN(46,1)/COMBIN(50,3) = 0.168980 \ (17\%)$$

This does not include a small additional chance of getting the other pair of the rank in your hand. Of course, every other player) gets the pair as well. There are many of such calculations (see Problems & Exercises for further examples).

Blackjack

No cards are exchanged in *Blackjack* but the player can ask for more cards to be revealed. Simple strategy in play is based on rough estimates of probability of conversion while avoiding being 'bust'. The value of your initial cards decides how many cards are available for a successful 'hit' within the limit of '21'. If you have a total of '12' there are more possibilities than if you have '19'. This in turn depends on how many cards are out or that you can see. On a casual level with a single deck in play, as cards are dealt out, note can be taken then used to gauge possibilities for conversion to a possible winning hand. For casino play with multiple decks, a record of played cards can be kept by more sophisticated *card counting* systems (see more below), of which there are a number (wizardofodds.com; Rosenthal).

Calculating a *player 'bust'* probability is relatively easy, at least for one additional card, e.g. you have a '5' and '7' – 'stand (stay) or hit (twist)?' Any '10' or a picture card will bust you and there are there are 16 of such cards and a general figure for a 'player bust on 12' is 16/52 = 0.307692 (31%). Of course, if we include the specifics of the particular play, with knowledge of your own cards, the dealer's up-card and any previous since the last shuffle, then this figure could be modified. Say, your hand and the up-card are not '10', but three have been seen in the previous hands along with another 12 cards.

No. of unseen cards = 52 -3 -3-12 = 34

No.10s remaining = 16 - 3 = 13

P(*'get 10 on next card'*) = 13/34 = 0.382353 (38%)

Doing calculations or simulations (not shown) for other player hand totals, shows, logically, that probability of a bust rises with the total, going over 50% at '14' and reaches 92% by '20' (www.blackjackage.com). This provides the basis for a strategy in that, on a relative basis, '12' has the lowest 'bust' chance, so 'always hit on 12 or less'. However, this is not the final ruling, and strategy for totals 13 - 17 are more problematical (www.lolblackjack.com).

We turn now to the *dealer's up card*, which refines the above decisions and once seen, a probability can be calculated before any other action. By this, whatever the up-card is, we can calculate the probability for the other card totalling up to 21 or bust. Thus, if the dealer shows a '10', the probability of a blackjack is,

P = 4/51 = 0.0784314 (8%)

There are 4 cards with a value of '11' and 51 cards are unseen.

Calculations for other up-cards are not so easy to do and may have to take into account several pathways and that the dealer stops when reaching a certain total. Thus, the rule may be, in private games at home, etc., that the dealer can act as the players, with no rule. In casinos, there are several variations of 'stand' rules, typically, that the dealer must hit on a '16 or less' and stand thereafter, even on a 'soft 17' (17 with an ace). The calculations reveal some unusual steps, as when the dealer card is '7' - to get this up to the first position where he/she stops ('17') would start as:

$$P('get\ 7\ to\ 17\ on\ 1\ card') = 16/51 = 0.313726$$

OR

$$P('get\ 7\ to\ 17\ on\ 2\ cards')$$
$$= (^4C_1)^2\ /2 + 3 * (^4C_1)^2 + {}^4C_1 * {}^3C_1\ /\ ^{51}C_2 = 0.0533333$$

OR ...

We can see that for the second event, to get '7 to 17', the dealer must get a total of '10' on 2 cards: '9+ace'/'8+2'/'7+3'/'6+4'/'5+5'. However, we cannot take these as a sum for a joint probability of '1 from 4' for all. First, there are only three '7s' available, and second - the ace must come second in the '9+1' combination, hence the probability for this event must be halved. Adding the probabilities for '17 on 1' and '17 on 2' gives,

$$P('7\ to\ 17\ with\ 1\ or\ 2\ cards\ hits') = 0.367059\ (37\%)$$

Calculations for 'get 17' on 3 cards, 4 cards and 5 cards follow, with more and more improbable events such as '7 with a 6 and four aces', so the figure above accounts for most of the probability.

Performing each calculation will produce a matrix of values covering all possible up-cards and the totals attained on drawing one or more cards up to a top limit of '21' (wizardofodds.com). Looking at such tables, players can see that the most 'dangerous' up-cards are the ace, and 7-10. Of course the dealer can 'bust' as well, rising from up-card of '2' (35%) to a maximum at '5' (43%), then falling from 26% at '7' to 21% at '10' (www.lolblackjack.com).

Bringing these together with further in-depth analysis, in particular, *expected values* for all combinations of player hand and dealer up-card, are used to formulate **basic strategy** in *Blackjack*, with the higher expectation deciding the issue. From the player's point of view, 'stand' is definite on '17 or more', hitting is usual on '12 or less' and possible on '12, 13-16', but this depends on the dealer's up-card. The higher bust probabilities for values '4, 5 and 6' (when dealer totals are under '17' and he/she must hit) can prompt a hit when these are combined with player '12 or less'. Thereafter, 'stand' is ok, up to when the dealer has an up-card of '7 or more', where he/she has better conversion probabilities with lower bust chances. Here, it is worth taking the risk of a hit. This outline, does not consider the full detail of the player's hand and other actions such as 'doubling down', 'splitting', 'insurance', etc.

Card counting can improve the basic strategy (note: counting is 'frowned upon' by casinos). If the player does engage in this, then once some hands have been played, the key feature to note is when the remaining cards in the deck become imbalanced. This means that if mostly low cards ('2s to 6s') are seen, it becomes 'rich' in high cards (value '10') and vice versa - if many high cards are out, the deck is rich in low cards. Players can use this to their advantage in several ways, e.g., with a '10 rich deck', if they are dealt a total of between '12 and 16' then the probability of '10' on a hit is higher, than if the deck were 'low rich'. Conversely, 'low rich' would mean you would less likely to go bust even if you had a '16'. A '10 rich' deck also means that *blackjack* probability is higher all round, but players get higher odds (3-2) than the dealer (1-1) does for this hand. A count of aces is useful too.

Of course, keeping count over a long session is much more demanding, at least on the surface. Shuffling the cards after each deal or using multiple decks of cards as is done in casinos will obviate card counting or make it much more difficult, respectively, and probability will have to be based on the players own cards or any others seen within the current deal. There is much more to these effects and options on strategy, based on these measures with basic blackjack and more complex casino versions with 'doubling', 'insurance', 'splits', etc.

22.2.3 Expected values in card games

These can be calculated for the return at various points in betting games. *Video Poker* (5-card with the option to exchange 1-5 cards) lends itself most easily to this, as there is basically a deal followed by improvement if desired and there are <u>fixed pay outs</u>. Consequently, there are fixed expected values for various play actions. This contrasts with player versus player games (see below), where the expected value varies depending not only on the action but on the pot size, which can change as the game progresses.

The pay outs in video poker are usually directly related to the hierarchy described above, e.g. for a bet of 1 unit you might get back, for a final hand of 'nothing' 0, 'one pair' 1, ('Js' min.), 'two pair' 2, '3oak' 3,'straight' 4, 'flush' 6, 'full house' 9, '4oak' 25, 'straight flush' 50 and 'royal flush' 250.

The version with 'Js' minimum requirement means that hands with apparently 'nothing' can still have potential. An initial hand with a single 'J' to 'A' can form 'Js or better' on exchange, or any card 'J or higher' is worth keeping. The ER for these single card cases also depends on the discarded cards - if you have 'A♥' and there are 2 '♥s' in the discards then there are less chance of a flush in '♥s' cf. a single 'Q♦' with no '♦s' in discards. Additionally, the 'Q' can make more straights than the 'A'.

You are playing video poker and the initial hand is '2♦,Q♦,6♦,4♠,5♠' (a **'3-flush, 4-inside straight, 3-open straight'**). What are the options to maximise the expected return?

The general procedure is the same as that used for the draw poker hand (Section 22.2.2), so some workings are minimised. The various options and probabilities are:

Option 1 -Hold three '♦s' and change 2 cards

$$P('3\ flush\ to\ 5\ flush') = {}^{10}C_2 / {}^{47}C_2 = {}^{45}/_{1081} = 0.0416281$$
$$P('3oak') = ({}^{3}C_2 + {}^{3}C_2 + {}^{3}C_2) / {}^{47}C_2 = {}^{9}/_{1081} = 0.00832562$$

$$P('two\ pair') = 3 * (^3C_1)^2 / {}^{47}C_2 = {}^{27}/_{1081} = 0.0249777$$

(two from 3 (3 ways) of '2,Q,6' from the remaining cards in these ranks (3))

'Js or better': this requires more calculation:

The two drawn cards can form a pair of 'Js, Ks or Aces' on their own: $^4C_2$ *3 =18, and a 'pair of Qs' requires a second 'Q 'and 1 odd card that can be drawn from ranks with 4 cards available (full ranks) or partial ranks (the '4s' and '5s' with 3 cards available),

$$= {}^3C_1 * ({}^8C_1 * {}^4C_1 + {}^2C_1 * {}^3C_1) = 114$$

$$P('Js\ or\ better') = (18 + 114) / {}^{47}C_2 = 0.122109$$

These probabilities sum to 0.197040 - a 20% chance of winning (at least as 'Js or better') and the complement gives the losing chance as, 1- 0.19704 = 0.80296, then,

$$ER = -1 * 0.802960 + 0.122109*1 + 0.0249777 * 2$$
$$+ 0.00832562 * 3 + 0.0416281 * 6$$
$$= -0.356151$$

The win expectations on their own sum to 0.446809, which is the usual style for reporting the return values.

Thus, like other gambling games video poker has a negative return - but is keeping the three '♦s' the best way to progress? Well, remembering the value of cards with rank above '10' an alternative is to retain the 'Q' and exchange 4 cards.

Option 2: Hold 'Q' and draw 4 cards

This action means that the hand can form all the hands of the previous choice plus full house, 4oak, straights, straight flushes and even a royal flush. The calculations for these are relatively simple but those for pairs, 3oak and two pairs involve use of full and partial ranks and mixtures of the two (these calculations are not shown but are similar to some in the draw poker hand, Section 22.2.2). The expected return gives,

$$ER = -1*0.66502957 + 0.00116054*6 + 0.00428896*4 + 0.2548482*1$$

$$+ 0.00029154*25 + 0.02299779*3 + 0.04975191*2 + 0.00161467*9$$

$$+ 0.0000112130 * 50 + 0.00000560648*250$$

$$= -0.19378$$

The winning chance is 0.334970, about 33% and this action loses less on average, with the win values summing to 0.471248, so this is the better option, so far.

Option 3: hold all except the 'Q' and try for a straight.

Trying to convert an 'inside straight' is generally not recommended. What can this route produce?

We need a '3' for the straight,

$$P(\text{'4 ins. str. to 5 str.'}) = {}^4C_1 / {}^{47}C_1 = 0.0851064 \ (9\%)$$

There is also chance of a pair but in this case, it would not be 'Js or better' and does not contribute to ER,

$$P(\text{'pair on 1 card'}) = {}^4C_1 * {}^3C_1 / {}^{47}C_1 = 0.255319 \ (26 \ \%)$$

$$ER = -1 * (1-0.085) + 4*0.085 + 0 * 0.255319 = -0.575 \ (0.34)$$

Option 3 gives the lowest return and overall option 2 has the best potential. In these calculations, if have a certainty then the pay out is added on without a probability multiplier, e.g. if you already have 3oak and change 2 cards, the expected return has the 3oak pay out added to it.

With other forms of poker, as in player-to-player, expected values can be calculated at each stage of betting, using the probability of making an improvement, balanced against the monetary gain. The same scenario could be interpreted differently depending on the bet amounts and pot odds – essentially the probability of success could be high on the exchange but the monetary gain may give a low return.

In *Texas Hold 'em*, bets occur at four stages and expected return can guide whether to call or to pull out (fold). Take a 2-handed game where the blind and pre-flop bets have given a pot of 10 units. A player has an initial hand as a pair of 'As', which pre-flop has the highest probability of a win (85% with 2 players). The player is required to call 5 units to stay in:

ER = -5 * 0.15 + 10*0.85 = 8.5 – 0.75 = + 8 approx.

Thus, a clear decision to call or even bet up (raise) to the pot or more.

Other pocket card combinations have less initial potential, e.g. a pair of '7s' has about 65% chance of winning and two off-suited cards with a wide gap give the poorest e.g. '2♣,7♥' around 30%:

ER = - 5 * 0.7 + 10*0.3 = 3 – 3.5 = - 0.5

A negative return, so in the long term this type of bet would be unadvisable (again, with poker, the number of opponents, their behaviour and your table position may prompt you to make a different decision).

Expected values become more important as the game progresses and the pot gets larger. In addition to your own hand, judgements on the opponent's hand (unseen) are vital. One tactic is to visualise the best that they could have and estimate your win chances.

Rather than work with probabilities, which are difficult to remember, poker players can use '*outs*' – the number of cards from the remainder to give you a win:

Probability of success on next card(s):

= no. outs / remaining unseen cards

These are most easily calculated on the basis of 'the next card to come'. In the case above, the simplest example is where you have 2 suited cards and there are also two of the same suit in the flop – you need 1 on the 'turn' card to make a flush. There are 9 cards in the suit remaining, thus, there are '9 outs' and P = 9/47 ≈ 20% (1 in 5). Contrast this with 2 pocket 'Ks' with no cards of use in the flop - on the turn you need a 'K' to make a set, therefore '2 outs' (P = 2/47 ≈ 4% (1 in 24)). The fewer 'outs' the lower the probability that you make it.

'Outs' can be calculated for more than one card but with more complexity. For example, before the turn you have 'J,10' and '9,Q' in the flop. To get a straight, you need a 'K' or '8' by the river and there are 4 of each unseen - '8 outs' in total. These are made up of 4 cards each time - 4 for the 'turn' and 4 for the 'river'. The required cards can be obtained as,

'K on turn, + any on river exc. 8' = 4/47 * 39/46 *2
OR
'8 on turn, + any on river exc. K' = 4/47 * 39/46 *2
OR
'K on turn, 8 on river' = 4/47 * 4/46
OR
'8 turn, K on river' = 4/47 * 4/46
OR
'two Ks' = 4/47* 3/46 OR 'two 8s' = 4/47* 3/46

$$P = 4/47 * 39/46 * 2 + 4/47 * 39/46 * 2 + 4/47 * 4/46$$
$$+ 4/47 * 4/46 + 4/47 * 3/46 + 4/47 * 3/46$$
$$= 0.314524 \ (31\%)$$

22.3 Other card problems

Some other card probabilities and questions are of interest within games or for possible game design. Take a simple betting issue such as *'I bet you that I get a jack on the 6th card!'* or *'within 6 cards!'* A key influence is whether or not the cards are returned during selection.

Average number of draws (without replacment) and success by the nth draw

These problems look for probabilities and the average success point by a specific draw when dealing, laying out cards or spreading the deck face up. Taking the exampled posed above, if this is done *with replacement* and shuffling in between (a rather tedious process), we can use can the geometric and negative binomial distributional models (Section 21.3). Thus, via the expected value of the geometric ($1/p$) it takes on average 13 cards for a particular rank to appear and 52 for a particular card. So, using the NEGBINOM.DIST( ) function, the bet above *'a J **on** the 6th card with replacement'* (5%) is more likely than on the average at the 13th (3%). Specifying *'by the 6th'*, cf. *'by the 13th'* switches the probabilities around with P = 31% and 38%, respectively.

The probability and expected value for the *without replacement* equivalent of such events are not easily attained. If we take a particular card first, as *'How many cards, on average, do I need to draw or deal out before the 'K*♣*'* *appears, without replacing the drawn cards?'* This cannot be done with the geometric, which requires replacement, and the binomial is for a fixed number of trials. This question is answered by using the structure of the geometric method, but modifying the formula for success (p) and failure (q) of the event, which change after each step, thus,

Spreadsheet Table 22.2

Trial no.	Success(without replacement)	Failure
1	1/52	51/52
2	1/51 * 51/52 = 1/52	51/52 * 50/51 = 50/52
3	1/50 * 51/52 * 50/51 = 1/52	51/52 * 50/51 * 49/50 = 49/52
...	...	...
10	1/52	42/52
52	1/52	0

The probability of success on the 'nth' draw is the product of 'n-1' fails followed by success, so success on the 4th would be,

$$P = {}^{51}/_{52} * {}^{50}/_{51} * {}^{49}/_{50} * {}^{1}/_{49} = (51 * 50 * 49) / (52 * 51 * 50) * {}^{1}/_{49}$$
$$= {}^{49}/_{52} * {}^{1}/_{49} = {}^{1}/_{52}$$

Looking at these figures, we can get a formula. The success part is simply

$1/(52-(n-1))$. For the fail, the denominator is a factorial from 51 to 51- n, and the numerator 52 to 52-n, hence,

P(*'success on nth draw, without replacement'*)

$$= 1/(52-(n-1)) * PERMUT(51,n-1) / PERMUT(52,n-1)$$

We can generalise this for different sizes of the original group or pool, and the particular subgroup, as:

P(success on nth draw, without replacement)

$$= 1*sg /(N -(n-1)) * PERMUT(N-sg, n -1) / PERMUT(N,n-1)$$

Where, N = size of pool, original group

n = trial where success occurs, not > N and (N-sg)

sg = size of subgroup within N

(For all cases where all objects are distinct, sg =1)   *(22.1)*

For fully named cards (22.1) gives $^1/_{52}$ for all positions in the deck (see example below). It is applied below for subgroups (and in some previous examples (Section 17.4)).

These are the probabilities of waiting until the nth trial, where n = 1 to 52, to achieve success. What is measured is the probability of failing n-1 times, followed by the success. Thus, to illustrate with another example, when dealing out cards and turning them face up, the probability of having to wait until the 10th card to get the '$K\clubsuit$' is, sg =1 n =10 N =52,

$$P = 1*1 /(52 -(10-1)) * PERMUT(52-1, 10 -1) / PERMUT(52,10 -1)$$
$$= 0.0192308 = {}^1/_{52}$$

The same probability applies to any trial from 1 to 52 inclusive, i.e. it remains constant, just slightly more than the 'with replacment' condition (0.0161472). These results should not be confused with the probabilities for drawing from a reduced deck, i.e. conditional probability – given that 9 cards are out – what is probability that next card (10th) is '$K\clubsuit$'? Here, p = $^1/_{43}$, the 40th card = $^1/_{13}$. If we state these as the probability that we will have to wait until the 10th / 40th draw, then both are $= \frac{1}{52}$.

For subgroups, calculations were prepared as done above and using (22.1) for ranks, with sg $= 4$, N $=52$:

P('*rank, success on nth draw without replacement*')

$= 1*4 /(52-(n-1))) * PERMUT(52-4, n-1) / PERMUT(52, n-1)$

This time, the fractions do not cancel out, resulting in different probabilities, which decrease as the draw progresses, e.g. the probability of success on the first trial is $\frac{1}{13}$ (0.076923, 8%) but having to wait until the 6th draw before a 'jack' appears is 0.0560712 (6%; see below). A similar treatment of calculations can be done for a particular suit, with sg=13 and,

P('*wait until 6th card for a spade*') $= 0.0612753$ (6%)

Therefore, rank and suit have approx. the same probability at 6th card, which could be a useful betting ploy (probability is greater for the suit up to the 6th selection then it lessens).

To get the **expected value** for 'without replacement', i.e. the average number of draws until success, we need to use the long method – summing until nth trial for the trial no. * probability,

E(X) 'wait until get K♣' $= 1 * \frac{1}{52} + 2 * \frac{1}{52} + 3 * \frac{1}{52} \ldots 52 * \frac{1}{52} = 26.5$

(We can shorten this if we turn to the method used previously, (formula (15.0) Section 15.2.1) and,

1 to 52 $= 1+2+3+\ldots 50+51+52 = (53*52)/2 =1378$ and

E(X) $= 1378 * \frac{1}{52} = \frac{53}{2} = 26.5$)

So on average, we would have to turn up 27 cards to get to the 'K♣' (cf. 52 *with replacment*).

The expected value for ranks and suits was obtained by enacting the formula for n $= 1$ to 49 (ranks) and 1-39 (suits) to produce a list of probabilities (not shown). E(X) is obtained via the sum of the product of each trial no. and its probability, as was done for the single distinct card example above. For ranks and suits, the SUMPRODUCT( ) function was applied to the list of n and p values mentioned previously,

> E(X) 'no. cards until a particular rank' = 10.6 (cf. 13 *with replacement*)
> E(X) 'no. cards until a particular suit' = 3.79

We can see that in the jack example it would take 11 cards on average to reveal this rank value.

Position in the deck and the hypergeometric function

The hypergeometric distribution covers the without replacment state and it can be applied to these problems (as shown but not detailed in Book One Section 13.3). The HYPGEOM.DIST( ) function is used in an unusual way for certain card probability puzzles and some others. With cards, as seen above, the problem regards the position of cards on the deal or within the deck, or when cards are spread out or fanned. All cards prior to the position of interest are viewed as being taken and retained, so probabilities are dependent as for the 'without replacement' state.

A single card can occupy one of the 52 locations in the deck. As we calcualted in the previous section, the probability that it is found on the top = $^1/_{52}$, bottom = $^1/_{52}$, 5th = $^1/_{52}$ etc. If this looks too easy, and you are not convinced we can look at this another way – it's rather like the geometric except without replacment, e.g. on the 5th there would be 4 fails then a success. We can apply the hypergeometric function to this, so with the problem, '*what is the probability that we select 5 and get the J♦ last (i.e. on the final card)?*',

> P = HYPGEOM.DIST(1,5,1,52,0) /5 = 0.01923 = $^1/_{52}$

The function gives the probability for *any order* but to get the orders where the 'J♦' is the last card, we need to divide by a factor, based on the arrangements possible. Assuming orders are equiprobable, this is calculated by following the reasoning for restricted permutations in Book One Section

7.4,

No. permutations of 4 cards all ending in 'J♦' = 4!

So, we multiply by this number and the net adjustment is,

Adjustment for probability = * 4!/5! = * $^1/_5$ or divide by 5.

Thus, all possible orders are formed from the 5 cards by 'w,x,y,z,(J♦)', which is = 5! but only a fifth of these will have the 'J' last and the function result is divided by 5 as above.

Applying this, we can confirm the examples above,

P('on the 6th for spade') = HYPGEOM.DIST(1,6,13,52,0) / 6 = 0.0612753

on the basis that the '♠' can be at 1 of 6 positions in the draw.

Similarly,

P('a Q on the 3rd')= HYPGEOM.DIST(1,3,4,52,0)/3 = 0.0680543

Thus, returning to our bet with 'a J on the 6th card' we can confirm the probability for this:

HYPGEOM.DIST (1,6,4,52,0)/6 = 0.0560712 (6%) 17 to 1 against

The other bet is 'within 6', so the 'J' could appear anywhere within a deal of 6,

HYPGEOM.DIST (1,6,4,52,0) = 0.336430 (34%) 2 to 1 against

and at least 1 'J',

1- HYPGEOM.DIST (1-1,6,4,52,1) = 0.397230 (40%) 3 to 2 against

Therefore, this is an altogether different bet and cannot be viewed as comparable. The average number of draws for a 'J' is 10.6 (≈11), which gives probability ('on the 11th'),

HYPGEOM.DIST (1,11,4,52,0)/11 = 0.0393758 (4%)

These probabilities are at the same relative level as the *with replacement* state, but slightly higher (on the 13th card, P = 0.0337575 (3.8%)). We can use the function to give other probabilities at various points,

P('*by the average (11th card)*') = 43%

P('*by the 13th card*') = 44%

Again, slightly higher probabilities and overall the bet would be more favourable for the bettor in the *without replacement* condition.

This application of the function can be expanded for other problems such as '*what's the probability that all the 8s are in the bottom half of the deck?*' Take the top half as up to card no. 26 – so we must have 'no 8s' by then:

P = HYPGEOM.DIST(0,26,4,52,0) = 0.0552221 (6%),

This says that there is a low probability that there are zero '8s' (i.e. none) in the first 26 cards. Via the complement, we can say that there is 94% chance that '1 or more 8s' are present.

In game play, e.g. after a deal out of 20 cards as four 5-card hands, this calculation gives the probability of '*zero aces*',

P = HYPGEOM.DIST(0,20,4,52,0) = 0.132829 (13%)

In addition, the chance of '*at least 1 ace*' is,

P = 1 - HYPGEOM.DIST(0,20,4,52,1) = 0.867171 (87%)

If we just want to know that a card is present in the draw or deal, we can omit the division by the factor as illustrated above and,

P('*J$\blacklozenge$ in first 5 cards*') = HYPGEOM.DIST(1,5,1,52,0) = 0.0961538 = $^5/_{52}$

Individual probability confirms this result:

P = $^1/_{52}$ * $^{51}/_{51}$ * $^{50}/_{50}$ * $^{49}/_{49}$ * $^{48}/_{48}$, for being first position,

OR $^{51}/_{52}$ * $^1/_{51}$ * ..., etc.

These cancel out for the five positions and become $5 * \frac{1}{52}$ as above.

This is a quite powerful tool – we can get the probability of any card(s) being at any point in the deck or within any range – within top 26 cards or, within the bottom 26. For a fully named card 'within the top 26,

$$P = \text{HYPGEOM.DIST}(1,26,1,52,0) = 0.5 = \frac{25}{52}$$

or 'not in the top 26'

$$P = \text{HYPGEOM.DIST}(0,26,1,52,0) = 0.5$$

Twenty cards are dealt out - *'what's the probability that 4 kings are out somewhere?'*

$$P = \text{HYPGEOM.DIST}(4,20,4,52,0) = 0.0178964$$

We can also get some scoring patterns for games - Probability of *'4 aces in your hand (5 cards)'* $= \text{HYPGEOM.DIST}(4,5,4,52,0) = 0.00001847$

Then, *'any 4 oak, draw poker'*

$$= \text{HYPGEOM.DIST}(4,5,4,52,0) * 13 = 0.000240096$$

In *Blackjack*, you have '5,6' and the dealer shows '7' – *'what is probability that a '10' card (picture or face 10) lies with the next 2 cards?'* There are 16 of such cards and 49 are unknown,

$$P = \text{HYPGEOM.DIST}(2,2,16,49,0) = 0.102041 \ (10\%)$$

At least 1,

$$P = 1 - \text{HYPGEOM.DIST}(1-1,2,16,49,1) = 0.551020 \ (55\%)$$

There's not much chance of both being '10' but there a good chance for 1 of them. The player has the advantage as she goes first. Unfortunately, the chance for 'non-10' cards is much more likely (45% and 90%). (For more detail, see *Blackjack*, Section 22.2.2).

Arrangements in deck and multiple cards at a position together

Looking at these problems in another way, we can ignore position per se and consider subgroups of cards located in the deck in formations when spread or fanned out. The subgroup can be of cards in a particular set in order or just simply 'together'. We can also look at such events as a *run* or *streak* (Book One Section 5.3.3).

We know that all possible arrangement amount to 52! We require the specific arrangements of the nominated cards. Remembering the saying *'as thick as thieves'*, *'what's the probability that after shuffling, all the knaves (the jacks) are side by side in one subgroup of 4?'*

This is essentially a *'run of all the jacks'*. This cannot be solved using the run formula described in Book One Chapter 5, as we have the 'without replacment' state (it's not like 52 coins, because when one jack appears the probability for the next changes). Individual probabilities must be ascertained.

Taking the jacks ('JJJJ') as indistinguishable, they form a single unit. When the deck is fanned out, this can be located at any of 49 positions where the run can slot in – at the start, end and 47 gaps in the remaining 48 cards. The probability for finding the jacks at any one of these, e.g. at the start of the fan, is only 1 position out of the 49 and the other 48 cards can be in 48! orders, so,

$$P = 1 * 48! / 52! = 0.000000153908 \quad (\equiv \tfrac{1}{52} * \tfrac{1}{51} * \tfrac{1}{50} * \tfrac{1}{49})$$

- the latter expression gives the probability of one 'J' after another at the start.

For all positions,

$$P = 49 * 1 * 48! / 52! = 0.00000754148$$

The knaves in a pack of cards are extremely unlikely to stay together. If distinguishable, they can be arranged in 4! orders,

$$P = 49 * 4! * 48! / 52! = 0.000180995$$

$$\equiv 1/52 * 1/51 * 1/50 * 1/49 * \text{FACT}(4) * 49$$

When identified, the jacks seem slightly friendlier to one another but still a long way from being bosom buddies!

The final probability puzzle

To finish we look at a game that is more thought provoking. It can be performed with various props, but it is simple and convenient with cards. It is known by several names and it is an example of the *Monty Hall Paradox* (based on a game show with three doors behind one of which is a big prize). In the card version, three cards, one designated as the winning card (a 'Q', but any card is suitable), are randomly mixed by the dealer, who then places them out in a row (*Find the Lady*, a game with 3 cards and a 'Q' can be used to name the current game, although it usually refers to a version of *3-card Monte*, a confidence scam). The player can place bets and must indicate the winning card, but not turn it over at this point. The dealer then shows (turns over) one of the other cards (it's not the 'Q') and asks if you wish to stay with your chosen card or switch to the other unrevealed card – what should you do to maximise chance of a win? What are the probabilities for 'stay' and 'switch'? If playing with friends for chips or tokens etc. place your bet on your card of choice.

The initial choice has probability of $^1/_3$ by use of the classical formula. One card is revealed (not the 'Q') and two remain. Applying the same logic, one would say that the probability is split 1:1 and $P = {}^1/_2$. This has been shown to be incorrect and there are reams of explanation and controversy on this puzzle because of its counter-intuitive nature (vos Savant). The simplest way to convince yourself is to do a simulation and compare the incidence of correct choice. To elucidate by calculation, drawing a tree diagram (Fig. 22.2) can help.

In the diagram, the 'Q' is located at A and B or C is revealed, depending on the player's choice.

The display covers the situation for the 'Q' at A - similar diagrams can be constructed for other locations. The net result is that 'stay' has 1 correct and 2 wrong, but 'switch' has 2 correct and 1 wrong. Thus, based on 'switch vs. stay', the 'switching' wins 2 out of 3 (P = 2/3), a higher probability cf. 1 out of 3 for 'stay'. Note that in performing this game, the dealer must be able to manipulate the cards convincingly for a random shuffle but retain knowledge of the 'Q's' location.

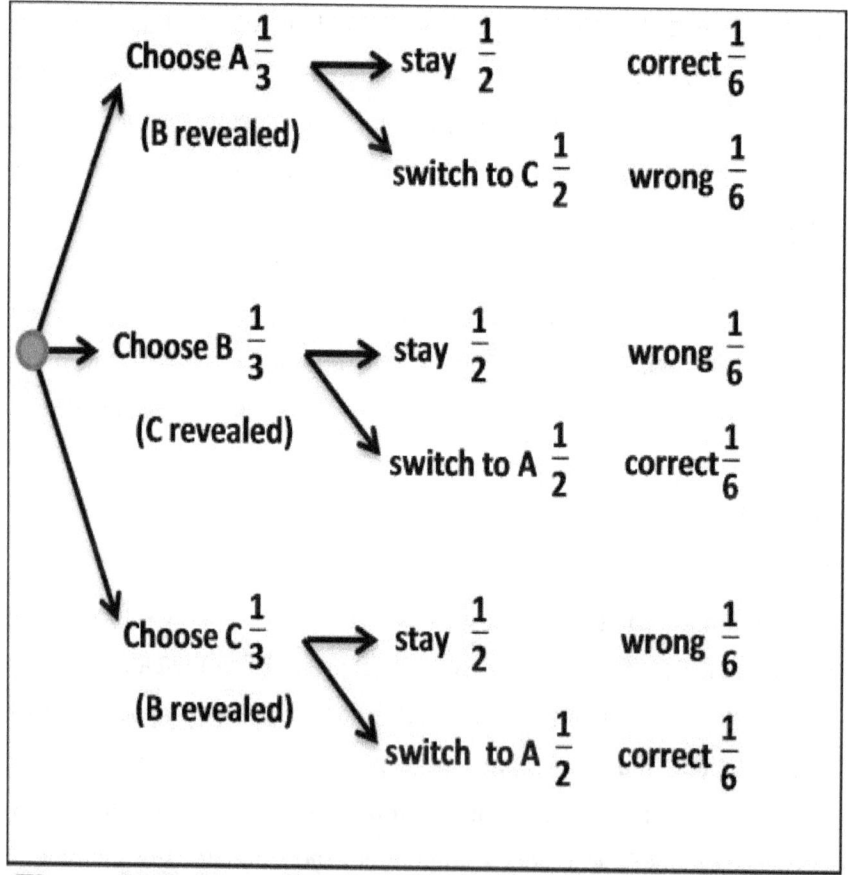

Figure 22.2. Tree diagram of choices in 'Find the Lady'

Additionally, the procedure above follows that of the original game show, where the prize (very expensive) was never revealed after the player had made the initial selection. If this is allowed, the dealer exposes the queen to show that the player has lost and no choice follows. This would modify the probabilities. Try out the game with friends where one player follows the 'always stay' strategy and another 'always switch' at a 1:1 bet for a win - you will soon see who has the more winnings.

We leave you to ponder this problem along with others that we have found along the way in our explorations of probability.

Problems & Exercises

1. Derive a formula in terms of the dice size ('s') for 'who wins' when 2 players roll dice (1 each) based on formula (15.0) in Section 15.2.1, i) for same size dice. ii) For mixed sizes. iii) For ties.

2. Two players, Andrew and Jeff, throw dice of the same and different sizes. Higher face value wins. Using the formula generated in Question 1, what is the probability that Andrew wins, when, i) both throw 1d12? ii) Andrew throws 1d4, Jeff 1d20? iii) Show that the probability for Jeff to win is the complement of this minus the ties. iv) Andrew uses 2d6 vs. Jeff with 1d6 (without using the formula).

3. Simultaneous duels - i) elucidate the formula for 'both hit'. ii) What is the probability for both hitting a target where a = 0.2 and b= 0.7 (contest ends when 1 or both hit)? iii) What is the probability that they both miss for 5 rounds? iv) If A is faster on the draw and has an 80% chance of beating B, what chance has A getting the drop on B and firing first to win?

4. What is the probability of getting a pair of 'Js', in three successive deals of *Video Poker* in a marathon session at the game (1000 deals)? How often would you expect such a streak to happen?

5. Daria the dart player needs to get '100' to finish a 501 in one turn (up to 3 throws). Her probability for singles, doubles and trebles are 0.8, 0.2 and 0.1, respectively, what is her chance for success if she goes for, i) Out in two throws - on treble 20 + double top? ii) Treble 20 + single 20 + double ten? iii) Double 20 + double 20 + double 10?

6. Careful study of relevant data and some subjective judgement convince a punter that team X can win the rugby championship cup.

He decides that they have a better than evens chance (52%) but on checking the bookmakers they have odds of '8/13' and another at '3/13'. What do these figures mean and what would he get back on a £10 bet by his odds and the bookmakers?

7. What's more probable - *'an even no. of Ts with 10 coins'* or *'an even no. with 6 coins'*? Can you figure out a short cut to reduce the number of calculations?

8. In the coin game (Section 16.6) with four players, Stacey, Nicole, Janice and Dane, flipping in order until the first H, calculate the probabilities for Nicole's, Janice's and Dane's chance of a win.

9. In *Scrabble®*, Pushka draws 7 letters out at the start of the game - what is the probability that she gets the word 'ENTERED' and gets to enter it to win a large score?

10. Calculate the expected value for the tile problem (the letter 'Q') in Section 17.3 for the *without replacement* procedure for, 1) the letter 'Q' ii) the letter 'E'.

11. Confirm the badge selection problem in Section 17.4 with individual probability.

12. In the Irish lottery, a claim is made that the jackpot odds are '42% better'. Is this so?

13. On week 1, a lottery draw has numbers 34,7,12,48,2 and 22. The next week the same numbers are drawn. What is the probability of this happening?

14. In the '322' raffle (Section 17.5.2), calculate i) the probability of winning all three prizes if you buy 60 tickets instead of 3. ii) How many would you need to buy to have greater than a 50% chance? If

you stick with 3 tickets what's your chances of iii) Winning nothing? iv) Winning at least 1?

15. One scratch card has 12 symbols and you are allowed to scratch all fields. You win if you get 3 or more matching symbols (3 = £3, 4 = £5 and 5 = £10. If the manufacturers produce 10% of all cards with 3oak, 5% with 4 and 1% with 5 matched sets, what is your probability of winning something on purchase of one card and what is the return if 5000 tickets are printed?

16. A scratch card has 16 symbols and you need to get at least 2 $)$s with any other (but not ☆s), or three different symbols. If there is a random placement of 4 $)$s, 10☆s and 2 ★s what is your chance of a win on scratching off 3 symbols?

17. '*What is the probability of getting 'two red numbers, two black numbers and two green occurrences', in 6 spins of the roulette wheel (American)?*'

18. Calculate i) the probability for '*three trebles and a single on 10d6*'. Which is higher ii) A treble on 5d6 or 6d6? and iii) 5oak on 7d6 and 9d6?

19. A selection of polyhedral and other dice are used to simulate planetary movement in a space exploration game, where 1d2 = mercury, 1d4 = Venus, 1d6 = Earth, 1d8 = Mars, 1d10 = Jupiter, 1d12 = Saturn, 1d14 = Uranus and 1d16 = Neptune. The span of years in the game for an eon is equal to the total number of sequences. If the 'conjunction' or rather alignment of two or more planets occurs when all dice are thrown and they all show the same face value, how often will this happen? Calculate the length of an

eon and the probability of all planets in alignment, and an example of one with Jupiter, Mars and Earth.

20. Using Table 18.2a and b, count the ways for C to win where A, B and C throw 1d6 - <u>highest roll</u> wins.

21. What's of higher probability, - cut the deck at a 'K' or get 'three Ks' at *Poker Dice*?

22. In *Pig*, i) what's the probability of finishing the game in one turn, assuming average rolls? ii) How many rolls need to be successful ('>1') before the probability falls below 50%?

23. Calculate probability for i) '4,x' on 2d6, where constraints are first: x $\neq$ 4 and second: none. ii) '7,7,7,7,x,y,z', x$\neq$y$\neq$z$\neq$7 on 7d10. iii) Probability of '3,3,5,5,5,w,x,y,z' on 9d6, w$\neq$x$\neq$y$\neq$z$\neq$3,5.

24. What's the E(X) (average number of throws) for i) *'any three of a kind with 3d6'*? ii) *'a double on 2d8'*?

25. Two players, Jono and Pushka, have a game of *Risk®*. In one battle, Jono throws a '5,6' on 2d6, then on the next attack throws the same. Pushka is disgruntled and loses a territory. How unusual are two '5,6s' in succession?

26. In a game, Jonathan says to Cassie, "Let's play dice for coins - exactly one '6' wins. I'll roll 5d6 but to give you an edge you get an extra die – 6d6 - but if I win, I get an extra coin". What are the true odds and does Cassie really have an advantage?

27. In a space game, missile targets must be hit at least three times for successful elimination. If the probability of a hit is 0.4, how many must be fired at each target to ensure (> 50% chance) of exactly and at least 3 hits each?

28. In a rpg, compare probability with 1d20 and 3d6 for i) getting an ability above '12' (where bonuses start). ii) Achieving a successful action ('15 or more' required) using an artefact with a -3 effect.

29. In a war game, an elimination is attained when a '5' or '6' appears on rolling a d6, a retreat with '4'. A 10 unit force is allowed to use 10 dice and requires to advance and eliminate a 7 unit force of similar ability. i) What is the probability of a complete victory (7 eliminations) or at least 7 retreats (assume a surprize attack and ignore defence factors, etc.)? ii) A typical game has 300 rolls for these units - how many eliminations would there be on average, assuming a similar balance of forces in each encounter?

30. In *Poker Dice*, calculate the probabilities for conversion improvement. You have a treble on the dice and you re-roll 2d6 - calculate all possible outcomes and confirm that the probabilities sum to unity.

31. In the game *Farkle*, determine probabilities for i) '*5 +any (except 1), one treble '*. ii) '*5 +any (except 1), one quad'*.

32. At a Halloween party with 10 people, what is the chance that at least 2 will have the same birth sign?

33. Edwin and Gordon play a board game and for 'who starts' the rules state: 'each player rolls 2d6 and the higher sum wins'. i) What is probability of a win for Ed? ii) What is the probability of a tie?

34. Calculate the probability of the sum of '21' on 2d14.

35. What is i) the probability for getting a sum of '9' on 5d4? ii) A sum of '33' on 5d8?

36. i) What would be the expected values for sum of the dice in Q.35 and

ii) What would be the expected no. of throws to get the above sums?

37. In a board game (*Escape from Colditz*® © Parker Bros., Hasbro Inc.), points for movement are generated by rolling 2d6, with any 'double' allowing a re-roll (added to the total) and so on. What is the expected sum of such a roll action?

38. In *Craps*, calculate the probability of converting the point values '6' and '5'.

39. i) Calculate sum = '200' on 50d6, 50d8 and 50d10, using normal approximation and compare with combinatorics (19.2a) using % relative error. ii) Calculate the 'at least 200' probabilities and explain the large difference between d6 and the other dice.

40. When would the probability (in cards) of drawing an ace equal 25%?

41. Confirm by the 'individual probability method' i) P('$J\clubsuit, 5\spadesuit$ *and* $9\blacklozenge$ *on deal of 7* ['$J\clubsuit, 5\spadesuit, 9\blacklozenge, w, x, y, z$'$J$']) in Section 21.3. ii) In Draw Poker, P(*'particular 4oak'*) iii) A straight in Draw Poker.

42. In *Texas Hold 'em* you have '$7\clubsuit, J\blacklozenge$' on the deal. Determine probabilities for getting on the flop: i) a match to get one pair. ii) A pair to give 3oak iii) a full house. iv) 3 cards for a straight. v) 3 cards to form 4oak.

43. In *Texas Hold 'em*, you have one 'Q' on the deal and there no usable cards in the flop and turn, but there is a 'K'. You require to put in 10 chips to stay in and the pot is at 50. i) What is your probability of getting another 'Q' on the river to get a high pair? ii) What is the risk in comparison with pot odds?

44. In a game of 7-card *Rummy* with no lay down until all melds are complete, Sandra plays against John. She has '$8\heartsuit, 9\heartsuit, 4\spadesuit, 7, 7, 7, 7\heartsuit$'.

What's the probability of Sandra getting out on the next card from the stock with 15 cards discarded (none usable)?

45. In a game of *Bridge,* you get dealt 2 hearts, 3 spades, 2 diamonds and 6 clubs. What is the probability of this hand? ii) Your partner has 3 clubs. What is the probability that your opponents have the remaining clubs as a '2-2' split?

46. In *Blackjack,* i) what is the probability of being dealt a '16' on the deal before any cards are seen? ii) Your hand is '10+6', the dealers up card is '8' and you decide to 'hit', what is the probability of a bust with 1 deck in play and multiple decks in casino play?

47. What is the probability of 'A♠' on the 20th card on a deal?

48. Compare probabilities in: i) *'Draw 3 cards and not get a '7' until the 3rd',* ii) *'Deal 3 and get one '7'',* iii) *'Deal 3 and get one '7' but not on the 3rd card'.*

49. *'After shuffling the deck, what is the probability of i) a 'nine' at position 1? ii) at position 9?'* iii) P('9' on 9th card'), irrespective of what cards 1 to 8 are?

50. i) On average, how many cards would you deal out before you see two spades? ii) What is the most probable number?

References

Alspach, B. (18 January 2000). '7-Card Poker Hands',
  http://people.math.sfu.ca/~alspach/comp20/

Bărboianu, C. (2009) *Probability Guide to Gambling*, Infarom Publishing,
  Craiova, Romania.

Batting average, http://en.wikipedia.org/wiki/Batting_average

Bayer, D and Diaconis, P. (1992) Trailing the dovetail shuffle to its lair, *Ann.
  Appl. Probab.* 2, no. 2, pp. 294-313.

Bingo probabilities, http://wizardofodds.com/games/bingo/probabilities/1/

Bingo probabilities, http://www.durangobill.com/Bingo.html

Bingo probabilities,
  http://www.nationalbingo.co.uk/natbingo.asp?pg=natbingo)

Bingo scoop, (Tuesday 04 November 2014),
  http://www.telegraph.co.uk/news/newstopics/howaboutthat/7038659/Bingo
  -player-scoops-18000-jackpot-in-record-time-on-first-ever-game.html

Blackjack, www.blackjackage.com

Blackjack, www.lolblackjack.com

Brad Mann, "How many times should you shuffle a deck of cards?"
  UMAP J. 15 (1994), no. 4, 303-332.

Bridge probabilities,
  http://www.bridgehands.com/P/Probability_Hand_Distribution.htm

Bridge probabilities, http://www.durangobill.com/BrSuitStats.html

Card counting, http://wizardofodds.com/games/blackjack/card-
  counting/introduction/

330

Casino poker dice, www.realmoneygames.org

Coin game ('2 Up'), http://en.wikipedia.org/wiki/Two-up

Collins, T. Probabilities in the Game of Monopoly®.
www.tkcs-collins.com/truman/monopoly/monopoly.shtml

Complete suit on deal, http://www.dailymail.co.uk/sciencetech/article-2065728/Whist-players-dealt-complete-suit-opening-hand.html

Craps record,
http://content.time.com/time/nation/article/0,8599,1901663,00.html

De Met, A. (2013). 'Dice Wars probabilities'. www.highprogramer.com

Dennis, S. (1999) 'Modeling Reality in Role-Playing Games: Insights from Mathematical Psychology'. *Places to Go, People to Be, The On-line Magazine for Roleplayers*, Issue 9, August 1999, http://ptgptb.org/0009/simon.html

Dice randomness test, http://www.awesomedice.com/blog/353/d20-dice-randomness-test-chessex-vs-gamescience/

Dice sums, http://www.superdan.net/download/DiceSamples.xls

Dice Wars probabilities, http://wizardofvegas.com/forum/questions-and-answers/math/7467-probabilities-in-dice-wars/

Dominated hands, http://wizardofodds.com/games/texas-hold-em/dominated-hand-probabilities/

Donato, M. (June 6 2008) 'Cracking The Horse Racing Code', http://bleacherreport.com/articles/27781-cracking-the-horse-racing-code

Draw Poker Odds Calculator, http://probability.infarom.ro/software.html

Epstein, R. A. (2009) *The Theory of Gambling and Statistical Logic*, 2nd ed., Academic press, Elsevier Inc., Burlington, USA.

Farkle strategy,
http://www.math.cmu.edu/~af1p/Teaching/OR2/Projects/P25/Farkle%20Re
port.docx

Football odds,
http://www.goal.com/en/news/2994/betting/2011/08/11/2615099/pete-
nordsteds-betting-guide-calculate-your-own-odds-to-find.

Gamescience, http://en.wikipedia.org/wiki/Gamescience

Garibaldi, J. *(2011)* 'Three Card Brag - the odds',
http://www.pagat.com/vying/brag.html#odds

Gelman, A. and Nolan, D. (2002). "Teacher's Corner: You Can Load a Die,
But You Can't Bias a Coin". *American Statistician* 56, 4, Nov. 2002: 308–
311.

'Go First Dice', http://www.ericharshbarger.org/dice/#gofirst_4d12

Gould, R. J. (2010) *Mathematics in Games, Sports and Gambling*, Chapman
& Hall, CRC Press, Boca Raton, USA.

Haigh, J. (2003) *Taking Chances,* (Oxford University Press), Oxford, UK.

Kern, J.C. (2006). *"Pig Data and Bayesian Inference on Multinomial
Probabilities". Journal of Statistics Education **14**
(3). http://www.amstat.org/publications/jse/v14n3/datasets.kern.html*

Lottery, https://www.national-lottery.co.uk/games/in-store/players-
guide/lotto

Lou Zocchi, http://en.wikipedia.org/wiki/Lou_Zocchi

McShane, J. M. and Ratliff, M. I. (2003) Dice Distributions Using
Combinatotrics, Recursion, and Generation Functions, The College
Mathermatics Journal, vol. 34,5 (Nov. 2003) pp. 370-376.

Mogensen[a], T. http://www.fantasylibrary.com/lounge/diceprop.htm

Mogensen[b], T. (2006) Dice-Rolling Mechanisms in RGPs,

332
http://www.darkshire.net/jhkim/rpg/systemdesign/torben_rpg_dice.pdf

Monopoly strategy, The Monopoly Nerds Blog. The strategy, tactics and Math behind Monopoly. www.monopolynerd.com

Mosteller[a], F. (1987) *Fifty Challenging Problems in Probability with Solutions,* Dover Publications, Inc., New York, USA.

Mosteller[b], F. (1987) "Isaac Newton Helps Samuel Pepys." Problem 19 in *Fifty Challenging Problems in Probability with Solutions.* New York: Dover, pp. 19 and 33-35, 1987.

Nishiyama, Y. and Humble, S. Winning odds (Penney Ante game), plus.maths.org

Packel, E. (2006) *The Mathematics of Games and Gambling*, The Mathematical Association of America, Washington USA.

Pass the Pigs, http://passpigs.tripod.com/prob.html

Risk game analysis, http://www.datagenetics.com/blog/november22011/index.

Risk strategy, http://www.totaldiplomacy.com/Home/tabid/67/CurrentPage/2/Default.aspx

Rosenthal, J. S. (2005), *Struck by Lightning*, Grant Books, London, UK.

RPG Design (2014) A Treatise on Different Dice-rolling Mechanics in RPGs, http://rpg-design.wikidot.com/evaluation

Scratch cards, https://www.national-lottery.co.uk/games/scratchcards/prizes

Shackleford, M. (29 may 2014) "Wizard of Odds- Slot machines explained".

Shuffling, http://russherman.com/Talks/CardShuffling.ppt

Shuffling, http://www.newscientist.com/blogs/onepercent/2011/07/shuffling.html

Slot machines, http://en.wikipedia.org/wiki/Slot_machine

Slot machines, http://entertainment.howstuffworks.com/slot-machine.htm

Spinner, http://commons.wikimedia.org/wiki/File:Monopoly_spinner.jpg

Stern, H. (1991) On the Probability of Winning a Football Game, *The American Statistician*, August 1991, 45, 3, pp.179-183

Su, Francis E., et al. "Seven Shuffles." *Math Fun Facts*. <http://www.math.hmc.edu/funfacts> .2500 shuffles

Texas Hold 'em odds, http://holdemtight.com/pgs/od/holdem-odds-page.htm

Texas Hold 'em, http://en.wikipedia.org/wiki/Poker_probability_Texas_hold_em

The Smack Down, http://scams.wikispaces.com/The+Smack+Down

'The World Record Craps Roll': Patrica DeMauro on May 23, 2009'. www.nextshooter.com. Retrieved December 28, 2013.

Tiddly wink records, http://www.etwa.org/records.html

Neller, T. W. and Presser, C.G.M. (2004) Optimal Play of the Dice Game Pig, *The UMAP Journal*, 25(1) (2004), pp. 25–47

Tote board, http://en.wikipedia.org/wiki/Tote_board

Townsend, N. (2014) The Sure Thing: The Greatest Coup in Horse Racing History, Century, London

Verhoeff, Tom (1999), *Optimal Solitaire Yahtzee Strategies*, Eindhoven University of Technology,

vos Savant, Marilyn (9 September 1990a). "Ask Marilyn". *Parade Magazine*: 16.

Walkenbach, J. (2010). *Excel 2010 Bible*, J. Wiley Publishing Inc., Indiana Indianapolis, p 31.

334

Weisstein, Eric W. "de Méré's Problem." From *MathWorld*--A Wolfram Web Resource. http://mathworld.wolfram.com/deMeresProblem.html

White Wolf games, http://www.white-wolf.com/

Winston, W. L. (2009) *Mathletics, How Gamblers, Managers, and Sports Enthusiasts Use Mathematics in Baseball, Basketball and Football*, Princeton University Press, Woodstock, UK

Yahtzee® strategy, http://www.bluffton.edu/mat/dept/seminar_docs/Elementary%20Farkle%20 talk.ppt

Yahtzee® strategy, www.yahtzee.org.uk)

Bibliography

Honary, E. (2007) *Total Diplomacy The Art of Winning Risk®,* BookSurge LLC, USA, www.totaldiplomacy.com

MacKinnon, R, F. (2010) *Bridge, Probability & Information*, Master Point Press, Toronto, Canada.

McLean, K.R. (1990) Dungeons, Dragons and Dice, The Mathematical Gazette, Oct 1990, 74, 469, pp. 243-256.

Olofsson, P. (2010) *Probabilities The Little Numbers That Rule Our Lives*, John Wiley and Sons, Inc., Hoboken, NJ, USA.

Phillips, T. (2008) *Beat the Odds*, The Infinite Ideas Company, Oxford, UK.

Tijms, H. (2009) *Understanding Probability*, 2nd ed. Cambridge University Press, Cambridge, UK.

http://thevirtuosi.blogspot.co.uk/2011/01/darts.html

http://www.frontiernet.net/~jamesstarlight/LinearVsNonLinear.html

http://reocities.com/ResearchTriangle/forum/2608/tdb1.html

http://datagenetics.com/blog/january12012/index.html

http://www.gamedesign.jp/flash/dice/dice.html

http://www.pinnaclesports.com/en/betting-articles/soccer/how-to-calculate-poisson-distribution

http://www.farklefun.com

Answers

1. The procedure in Spreadsheet Table 15.1 can be followed for these problems, but - i) The expression for summing integers (15.0), is converted by substituting X = 's', giving $((s-1) * (1+ s-1)/2) = (s^2 - s)/2$. This is divided by the number of sequences to get the probability of a win for one player, $P = ((s^2 - s)/2) / s^2$, (which reduces even further to $P = 1/2 - 1/(2*s)$ for same size dice. ii) For mixed dice the formula is used with 's' = smaller die, to give probability for the player with the smaller die to win. iii) For the probability of a tie, player 1's die = player 2's, i.e. a pair, the number of pairs is given by 's', again the smaller die is used for mixed dice and $P('tie') = s/s^2 = 1/s$ (same size) and $P = s_{smaller\ die} / s_{smaller\ die} * s_{larger\ die} = 1/s_{larger\ die}$ (mixed dice).

2. i) For 2d12, $P('Andrew\ wins') = [(12^2 - 12)/2]/12^2 = 66/144$, or more simply, $P = 1/2 - 1/24 = 11/24 = 0.458333$ (46%). ii) With mixed sizes, only the smaller die is used in the main part of the formula. For d4 vs. d20, $P('Andrew, d4\ win') = [(4^2-4)/2]/(4*20) = 6/80 = 0.075$. iii) Ties with 'd4 vs. d20' are based on d4, but with the larger die in the denominator, $P('tie') = 1/20 = 4/80$ and $P('Jeff, d20\ win') = 1 - 4/80 - 6/80 = 70/80$. iv) Higher face value wins are evaluated in *Risk®*, Section 20.2.2, where the probability of at least one face value beats each of 1d6 outcomes and are calculated using the binomial function. $P('Andrew, 2d6\ win') = 125/216 = 0.578704$ (58%).

3. i) both hit: follow the procedure in Spreadsheet Tables 15.2, 15,3 to derive $P = a*b /(b+a -a*b)$ as follows, Events are: E1 'A hit B hit' = $a*b$, E2 'A miss B miss then A hit B hit' = $(1-a)*(1-b)* a*b$, E3 'A

miss B miss A miss B miss A hit B hit', ... ∞. The first term is a*b

and the common ratio is (1-a)(1-b). Thus, P(*'both hit'*) = a*b/ (1-(1-

a)*(1-b)) = a*b/ (1 -1 +b +a -a*b) = a*b/(b+a -a*b). ii) P(*'both hit'*)

= 0.2 * 0.7/(0.2+0.7- 0.2*0.7) = 0.184212 (18%). iii) Using similar

procedures to i) the formula for 'both miss' is P = 1/(b+a -a*b) - 1.

With the data of ii) this gives P = 0.315789 (32%). This compound

event occurs 5 times in a row, P = 0.315789 ^5 = 0.00314042

(0.3%). iv) If A wins the draw (0.8), he/she fires first with overall

probability of P = 0.8 * 0.2/ (0.9-0.2*0.7) = 0.210526 (21%). Still

low, but B has to beat A to the draw and hit with P = 0.2 * 0.7/ (0.9-

0.7*0.2) = 0.184211(18%) so the gap has narrowed.

4. For each separate deal the event of a specific pair, $P = {}^4C_2 * {}^{12}C_3 *$
   $({}^4C_1)^3 /{}^{52}C_5 = 0.0325053$; isolated streak = three in a row, P =
   0.0325053^3 = 0.0000343450. No. opportunities = 1+ (1000-3) * (1-
   0.0325053) = 965.5922061; lambda (λ) = 965.5922061 *
   0.0000343450 = 0.033163219. P(*'one occurrence'*) =
   POISSON.DIST(1,0.033163219,0) = 0.032081457 (3.2%; 3.3%
   for 1 or more) (29 to 1 against, i.e. 1 in every 30 deals, so over 1000
   approx. 30 appearances of 'a pair of Js').

5. i) P(*'60+40'*) = 0.1*0.2 = 0.02 (2%). ii) P(*'60-20-20'*) = 0.1* 0.8* 0.2
   = 0.016 (1.6%). iii) P(*'40-40-20'*) = 0.2 * 0.2 * 0.2 = 0.008 (0.8%)
   (this ignores changing shots if misses occur for the original target,
   favourite doubles, etc.). Option i) has highest probability for Daria -
   although it involves a treble, there are only two throws.

6. The punter's estimate (say 52% of a win) would be odds of '12 to 13'
   against, (12/13), £12 for every £13 bet (92p per £1), giving £9.20p +
   stake = £19.20p. The bookies are not willing to go as high as this and
   feel that there is more chance of team X winning. One offers '8/13' (8

to 13 against) at $^{13}/_{21}$ (62%) chance of a win ($£^8/_{13}$ per £1 bet giving £6.20 + £10). The other is even more conservative and 3/13 (3 to 13 against, ($^{13}/_{16}$, 81% win)) gives only $£^3/_{13}$ per £1 bet = £2.30 + £10.

7. '0,2,4,6, 8 and 10' (or '0,2,4,6') can be calculated separately using formula (16.3a) or **BINOM.DIST( )**, (include 0 as even), then these are added; however, because the distribution is symmetrical '0,2,4' *2 (or '0, 2' *2) are equivalent, giving P('*even no. Ts*') = 0.5 for both.

8. Nicole = $^1/_2 * ^1/_2 + 1/_{16} * ^1/_4 + ...$ ∞. P('*Nicole win*') = $^1/_4 / (1- ^1/_{16})$ = $^4/_{15}$. Similarly, Janice's chance = $^2/_{15}$ and Dane's = $^1/_{15}$, so Stacey ($^8/_{15}$) has the largest advantage.

9. Assuming 100 letters of which E = 12, D = 4 and the other letters N, R, T = 6 each, using the multi-hypergeometric function (11.2b) gives P = $^{12}C_3 * ^4C_1 * ^6C_1{}^\wedge 3 / ^{100}C_7$ = 0.0000118744. A very rare event for Pushka.

10. i) Probabilities for 'getting the letter Q' are calculated in a similar manner to that for 'getting' a named card' on drawing *without replacement* (Section 22.3) as there is but one 'Q' in the mix. The fractions cancel out and probability for each stage is 1/100. Hence, E(X) 'Q' = 1 * 1/100 + 2*1/100 ...+ 100*1/100. Using (15.0) the sum of 1-100 = 5050, then E(X) = 5050/100 = 50.5. Thus, you would have to pick out 51 tiles on average (contrasting with the *with replacement* figure at 100 selections). ii) In contrast, the letter 'E' (12 occurrences), on average requires only 8 selections, without replacement (formula (22.1) used with subgroup of 12 out of 100).

11. P = 13/25 * 5/24 * 12/23 * 7/22 * 11/21 * 4/20 * 6/19 * 10/18 * 5/17 * 3/16 * 9/15 * 8/14 = 0.00000624964, which agrees.

12. Irish Lotto is a 6/45 lottery and the jackpot requires a six number

match, so $P = {}^6C_6/{}^{45}C_6 = 0.000000122774 = 1/8145060$. If you divide the UK national lottery jackpot probability by this, $= 0.0000000715112 / 0.000000122774 = 0.582462$, so its 58% less. Alternatively, if you calculate $0.000000122774 * 0.42$, you get the UK probability, i.e. the Irish Lotto jackpot chance is 42% higher.

13. $P = P(\text{1st draw}) * P(\text{2nd draw})$ but this is $\neq 1/{}^{49}C_6 * 1/{}^{49}C_6$. It does not matter what the first draw is - it can be any combination and it has $P = 1$. The second draw must match the first so P('*two consecutive lottery draws the same*') $= 1 * 1/{}^{49}C_6$.

14. i) P('*win 3, 60 tickets*') = HYPGEOM.DIST(3,60,3,322,0) = 0.00620755, so still below 1%. ii) You would need to buy 256 tickets to have a better than evens chance, P('*win 3, 256 tickets*') = HYPGEOM.DIST(3,256,3,322,0) = 0.501307 (50.1%). iii) P('*no win on 1st three draws*') $= {}^{319}/_{322} * {}^{318}/_{321} * {}^{317}/_{320} = 0.972224$ (97%) by individual probability, or by HYPGEOM.DIST(0,3,3,322,0). iv) P('*win at least 1*') $= 1 - 0.972223653 = 0.0277764$ (3%).

15. i) At start of sales for the particular batch, your probability of getting a win would be 10% OR 5% OR 1% = 16%. ER $= -1 * 0.84 + 3 * 0.1 + 5 * 0.05 + 10 * 0.01 = -0.19$, so a negative gain, you lose about 20p for every £1 spent (+0.65).

16. $P = ({}^4C_3/{}^{16}C_3) + ({}^4C_2 * {}^2C_1)/{}^{16}C_3) + ({}^4C_1 * {}^{10}C_1 * {}^2C_1/{}^{16}C_3) = 0.171429$ (17%).

17. Here there are three outcomes of interest: red (18/38), black (18/38) and green (2/38). Each spin can have one of these and the probability can be obtained by applying the multinomial function, $P =$ MULTINOMIAL(2,2,2) $* (18/38)^2 * (18/38)^2 * (2/38)^2 = 0.0125514$ (1%)

18. Use the MN calculator - i) P(*'three trebles, one single, 10d6'*) = 0.01667048 (2%). ii) P(*'3oak, 5d6'*) = 0.154321 and P(*'3oak, 6d6*) = 0.154321, i.e. they have same probability (15%). iii) Similarly for 7d6 and 9d6, both P(*'5oak'*) = 0.00900206 (0.9%).

19. Eon = 2*4*6*8*10*12*14*16 = 10,321,920, just over 10 million years. All planets in alignment would be when each die = '1' or '2' = 1 / 10321920 *2 = 0.000000193762, approx. once every 5 million years. For Earth, Mars and Jupiter we can have as '1s','2s','3s','4s','5s' or '6s' and the simplest specific case is when the three dice 'equal 1' and all the others are 'not equal to 1 and all different' = '1,1,1, m, v, s, u, n' = 1/(2 * 4 * 6 * 8 *10 * 12 * 14 *16) * 1 * 2 * 9 * 10 * 11 * $^8C_3$ * 3!/3! = 0.0107422, about once every 100 years.

20. Envisage C in last position of the throws in a row. Any sequence where the 3rd element is greater than the other two is a win for C. There are 120 singles (all values distinct), 20 of these end in a '6'; of the 20 that end in '5', 4*3 =12 qualify and similarly for '4s' (6) and '3s' (2), giving 20+12+6+2 = 40. There are 90 doubles, 15 of each value, and C wins when the odd value exceeds the pair, so all pairs '1 to 5' with an odd '6' (5), all pairs '1 to 4' with an odd '5' (4), '1 to 3' (3), '1 to 2' (2) and '1 to 1' (1) = 15. Total = 55 and P = 55/216 = 0.254630 (25%). Note: This problem differs from the example in Section 18.3.4, which has a specific target for the win roll.

21. P(*'cut at K'*) = 4/52 = 0.0769231 (8%); P(*'exactly three Ks with poker dice'*) = BINOM.DIST(3,5,1/6,0) = 0.0321502 (3%) (1 or more 'Ks' is much more probable (60%)).

22. i) 1d6 rolls where outcomes '2 to 6' are considered, have an average of 4 (= ((6+1) * 6)/2 -1) /5), then for 25 rolls of '4' to get 100 points, P = $^1/_6$ ^{25} - very remote. ii) P(*'3 rolls > 1'*) = $(^5/_6)^3$ = 0.578704

(58%), 4 rolls = 0.482253 (48%) and 17 rolls = 0.0450732 (5%).

23. i) both distinct by direct calculation, formula (18.9b), P = 1/6 * 5/6 * $^2C_1$ * 1! = 10/36; for no constraints, x can equal '4' so there is a repeated case - 1 instance '4,4', overall P = 11/36 = 0.305556 (31%).
ii) By direct calculation, P = $1/10^4$ * 9/10 * 8/10 * 7/10 * $^7C_4$ * 4!/4! = 0.00176400 (0.2%), which agrees with (18.10); the multinomial calculation gives the general case probability (any 4 of a kind on 7d10 and if divided by 10 gives the latter value. iii) all unnamed distinct (1,2,4,6), P = $(1/6^5)$ * 4/6 * 3/6 * 2/6 * 1/6 * $^9C_5$ * 5!/(2!3!) = 0.003000686 (0.3%).

24. i) P('3d6,three of a kind') = $^1/_{36}$, E(X) = $1/(^1/_{36})$ = 36. ii) P('getting a double on 2d8') = $^8/_{64}$, E(X) = $1/(^1/_8)$ = 8.

25. Each throw gives a fully named outcome, P = 1/6^2 * 2! Doing this twice, P('exactly 5,6, twice') = 2/36 * 2/36 = 0.0030864 (0.3%). Not a common event but there can be many 2d6 rolls in a full game.

26. For the throws, P('exactly one 6, 6d6') = BINOM.DIST(1,6,1/6,0), P('exactly one 6, 5d6') = BINOM.DIST(1,5,1/6,0) and both equal 0.401878 (40%) so odds are 1.5 to 1 against, or 2:3 in favour, for both players. However, in terms of expected return, Jonathan's = -1 * (Cassie wins = 0.4 * he loses 0.6) + 2 * (he wins, Cassie loses) = + 0.24, whereas Cassie's ER = -1 * (Jonathan wins = 0.4 * she loses 0.6) + 1 * (she wins, Jonathan loses) = 0, so Jonathan has the advantage.

27. According to the binomial calculation, 1-BINOM.DIST(3-1,7,0.4,1), 7 missiles will strike 'at least 3 times', with probability = 0.580096 (58%).

28. i) On d20, P ('13 or more') = 20-13+1 /20 = 40%, on 3d6 using

formula (19.2a) P = 26%, so on this comparison in the latter scheme, bonus abilities are less probable. ii) d20 P($'18$ or more$'$) = 20-18+1 = 3/20 = 15%, on 3d6, P($'18'$) = 0.5%, so a large difference, with the 3d6 roll success rare. While these results show the differences, individual systems can set the probabilities as wished within the games by adjusting difficulty targets and bonus/ penalty levels.

29. i) P($'eliminate'$) = BINOM.DIST(7,10,2/6,0) = 0.0162577 (2%) and P($'at$ least retreat$'$) = BINOM.DIST(7,10,3/6,0) = 0.117188 (12%), so more probable (at least 5 retreats have above 0.5 probability (62%)); (with a superior force, 10 against 7, 'at least 6 hits' can be calculated, signifying a possibility that some units get hit more than once, P = 0.0196616 (2%)). ii) Probability on a single elimination = 2/6 and on 300 trials, you would expect 300 * 2/6 = 100 successes.

30. Treble to 4oak, P = $^1/_6$ * $^5/_6$ * 2! = $^{10}/_{36}$ = 0.277778 (28%); treble to full house, P = $^1/_6$ * $^1/_6$ * 2! /2! *5 = $^5/_{36}$ = 0.138889 (14%); treble to 5oak, P = $^1/_6$ * $^1/_6$ * 2! /2! = $^1/_{36}$ = 0.0277778 (3%); retain 3oak, P = $^5/_6$ * $^4/_6$ = $^{20}/_{36}$ = 0.555556 (56%); add probabilities, $^{10}/_{36}$ + $^5/_{36}$ + $^1/_{36}$ + $^{20}/_{36}$ = $^{36}/_{36}$ = 1.

31. i) P($'6d6,5$ +any (except 1), one treble $'$) e.g. '5,2,4,6,6,6' = $^1/_6{}^6$ * $^4C_3$ * $^3C_1$ * 6!/3! = 4 * 3 * 120/6^6 = 1440 /6^6 = 0.0308642 (3%). ii) P($'6d6,5$ +any (except 1), one quad $'$) e.g. '5,4,3,3,3,3' = $^1/_6{}^6$ * $^4C_2$ * $^2C_1$ * 6!/4! = 6 * 2 * 30/6^6 = 360 /6^6 = 0.00771605 (1%).

32. Following the procedure described in Section 20.4 with the birthday dice, apply the same treatment for 'birth-sign' dice in the form of d12s. The probability is equivalent to getting at least 1 double, so we calculate the probability of none of the 10 having the same birth sign, i.e. 10 different face values on 10d12, then P = 1- PERMUT(12,10)/12^10 = 0.996132 (99.6%), so it's almost certain

(with 5 people its 62%).

33. i) Compare sums for '2d6 vs. 2d6' in the style used for *Dice Wars* (Section 20.3). The exact sum sequence counts for one player are multiplied by the sequence counts for all sums that can exceed them. These products are summed and expressed as a probability for the total of all dice used, in this case 4d6. $P = 575/1296 = 0.443673$ (44%), ii) Tied sums are given by the number of sequences squared, e.g. the sum of '8' has 5 sequences, so 25 in total for the event '8,8'. $P('tie') = 146/1296 = 0.112654$ (11%)

34. Use formula (19.1c) for the triangular distribution. Mode for d14 $= 15*2/2 = 15$, thus $k >$ mode and $P = (28 +1 -21)/14^2 = 8/196 = 0.0408163$ (4%).

35. i) Using (19.2a), $j = INT((9 -5)/ 4) = 1$, so the expression is evaluated twice. With $j = 0$, count $= 1 * {}^5C_0 * {}^8C_4 = 70$ and with $j=1$ count $= -1 * {}^5C_1 * {}^4C_4 = - 5$; Net count $= 65$ and $P('sum = 9, 5d4') = 65/ 4^5 = 0.0634766$ (6%). ii) $j = 33-5 / 8$, would give $j = 3$, but the lower sum can be used to give a lower j. The mode $= 9*n/2 = 22.5$ and lower sum $= 2* 22.5 - 33 = 45-33 = 12$, $j = INT((12-5)/8) = 0$ and $P = {}^{11}C_4 / 8^5 = 0.0100708$ (1%).

36. i) based on summing $E(X)$ for single dice, $E(X)$ '5d4' $= 5 * 2.5 = 12.5$ and $E(X)$ '5d8' $= 5 * 4.5 = 22.5$. ii) Based on the geometric mean, (trials till 1 success), $E(X) = 1/p$, so for 5d4 $= 1/0.0634766 = 15.753846$, 5d8 $= 1/ 0.0100708 = 99.296970$, so approx. 16 and 99, respectively.

37. The average sum on 2d6, based on doubles $= (2+4+6+8+10+12)/6 = {}^{42}/_6 = 7$. The series terminates when a non-double is rolled, $P = {}^{30}/_{36}$, so the average roll for these is the geometric $E(X) = 1/p = 1/{}^{30}/_{36} = {}^{36}/_{30} = 1.2$. Therefore, there are 1.2 rolls on average and the score is 7

* 1.2 = 8.4.

38. For '6' there are 5 ways to get '6' on 2d6 and 25 allow continued rolling (2,3,4,5,8-12) and 6 stop the process. Conversion goes as (5/36) OR (25/36 * 5/36) OR (25/36 * 25/36 * 5/36) ... ∞, a geometric series with 1st term = 5/36 and common ratio 25/36. Using (10.2b) gives P = 5/36 / (1-25/36) = 0.454545 and the win probability with '6' as the point = 5/36 * 0.454545 = 0.0631313 (6%). Point = '5' is treated in a similar manner. The initial probability is 4/36 and the geometric series sums to (4/36)/(1 - 26/36) = 0.400000, which gives 4/36 * 0.4 = 0.044444 (4%) for the win probability.

39. i) Average and standard deviation were calculated based on E(X) and V(X) for single d6, d8 and d10, then using approximation, for P('sum = 200, 50d6') = NORM.DIST(200+0.5,175,12.0761473,1) - NORM.DIST(200=0.5,175,12.0761473,1) = 0.00387934. Similarly for 50d8 = 0.00748899 and 50d10 = 0.0000215094. Combinatorics with formula (19.2a) gives 0.00389386, 0.00753116 and 0.0000191315, respectively. Assuming the latter figures are accurate, rel. errors (formula (14.1)) with the approximation are - 0.4%, -0.6% and +12.4%, respectively, error increasing with dice size. ii) Using 1 - NORM.DIST (200-0.5,ave.,SD,1) for the 'at least' probabilities gives 0.0212397 (21%), 0.942244 (94%) and 0.999899 (99.99%) for d6, d8 and d10, respectively. In the case of 50d6 (mode = 175), '200 or more' occupies a much smaller portion of the distribution area than that for the other dice, where their modes are above 200 (50d8 = 225, 50d10 = 275).

40. When dealing cards *without replacement*, if no aces have appeared by the time there are 16 cards left, the probability of the next card being an ace is given by 4/16 = 0.25 (25%) (many similar instances

can be found).

41. i) P('J♣,5♠ and 9♦ on deal of 7 ['J♣,5♠,9♦,w,x,y,z']') = 1/52 * 1/51 * 1/50 * 49/49 * 48/48 * 47/47 * $^7C_3$* 3! = 0.00158371. ii) P('particular 4oak') = 4/52 * 3/51 * 2/50 * 1/49 * 48/48 * 5!/(4!1!) = 0.0000184689. Alternatively, for a specific case with fully named cards, P('four 9s on 5 cards') = '9♣,9♠,9♦,9♥,x' = 1/52 * 1/51 * 1/50 * 1/49 * 48/48 * $^5C_4$ * 4!. iii) P('straight') = 40/52 * 4/51 * 4/50 * 4/49 * 4/48 * 5! = 0.00394004 (1st card must be 2-10 or ace = 40) this requires adjustment for straight flush = - 40 / $^{52}C_5$ = 0.00392465.

42. i) We need one '7' OR one 'J' in the flop but remember to subtract probability for 'both', which would give two pair, P = ($^3C_1$ * $^{44}C_2$ + $^3C_1$ * $^{44}C_2$ - $^3C_1$ * $^{44}C_{1)}$ /$^{50}C_3$ = 0.269388 (27%). ii)Two '7s' OR two 'Js' - we can't get 'both' this time but there is a possibility of the three flop cards being the same rank, so P = ($^3C_2$ * $^{44}C_1$ + $^3C_2$ * $^{44}C_{1}$ + $^{11}C_1$ * $^4C_3$) /$^{50}C_3$ = 0.0157143 (1.6%). iii) 'Two 7s plus a J' OR 'two Js plus a 7' are required, P = ($^3C_2$ * $^3C_1$ + $^3C_1$ * $^3C_2$) /$^{50}C_3$ = 0.000918367 (0.09%). iv) We need one each of '8s,9s,Ts', = $^4C_1$ ^3 /$^{50}C_3$ = 0.00326531 (0.33%). v) Either 'three 7s' OR 'three Js', P = ($^3C_3$ + $^3C_3$) /$^{50}C_3$ = 0.000102041 (0.01%).

43. i) There are 6 cards seen so 46 unknown and there are three 'Qs' outstanding, so you have 3 'outs' and P = 3/46 = 0.0652174 (7%, 14 to 1 against). With only three 'outs', this is a low chance and the danger is that your opponent could get another 'K' on the river, also with 3 'outs' (assuming opponent hand does not have a 'K' already). ii) The expected return on the next step would be ER = -10 * 43/46 + 50 * 3/46 = - 6.0869665, highly negative! The pot odds are 5 to 1 - much less than your chance of getting another 'Q'. There would have to be a much bigger pot to risk this action (about three times more in

the pot).

44. One of two cards would suffice '6♥' or '10♥' and P = 2/30 = 0.066667 (7%).

45. i) P = COMBIN(13,2) * COMBIN(13,3) * COMBIN(13,2) *COMBIN(13,6)/COMBIN(52,13) = 0.00470207 (0.5%). ii) P = COMBIN(4,2) * COMBIN(22,11)/COMBIN(26,13) = 0.406957 (41%) (more likely is '3-1' at 50%).

46. i) '16' by 10+6, 11+5, 9 +7, 8+8, P = $(^{16}C_1 * {}^4C_1 + {}^4C_1 * {}^4C_1 + {}^4C_1 * {}^4C_1 + {}^4C_2)/{}^{52}C_2$ = 0.0769231 (8%). ii) For casino play with say 6 decks, we assume that 52 cards are available at all times. Any card of rank value 6 or more on the next card will cause a bust -16 cards of rank value '10' and 4 each of '6,7,8, and 9', which give another 16 and P = 32/52 = 0.615385 (62%); with a single deck and knowledge of three cards: '10, 6 and 8', there are now 29 cards in the remaining unseen cards (49) and P = 29/49 = 0.591837 (59%; this is the approx. complement of the conversion probability for the blackjack example in Table 22.3 with slightly different circumstances).

47. As this is a fully named card, according to Section 22.3, P = 1/52, confimed by = HYPGEOM.DIST(1,20,1,52,0)/20 = 0.0192308 (and by formula (22.1).

48. i) P = fail * fail * hit = $^{48}/_{52} * {}^{47}/_{51} * {}^4/_{50}$ = 0.0680543 (7%) or HYPGEOM.DIST(1,3,4,52,0) /3. ii) P = $^4/_{52} * {}^{48}/_{51} * {}^{47}/_{50}$ OR $^{48}/_{52} * {}^4/_{51} * {}^{47}/_{50}$ OR $^{48}/_{52} * {}^{47}/_{51} * {}^4/_{50}$ = 0.204163 (20%), or more succinctly, = HYPGEOM.DIST(1,3,4,52,0)). iii) P = $^4/_{52} * {}^{48}/_{51} * {}^{47}/_{50}$ OR $^{48}/_{52} * {}^4/_{51} * {}^{47}/_{50}$ = 0.136109 (14%) or 2 * HYPGEOM.DIST(1,3,4,52,0)/3.

49. i) P('9' on 1st card'), = $^4/_{52}$ = HYPGEOM.DIST(1,1,4,52,0) =

0.0769231 (8%). ii) P('9' on 9th card) = $^{48}/_{52}$ * $^{47}/_{51}$ * $^{46}/_{50}$ * $^{45}/_{49}$ * $^{44}/_{48}$

* $^{43}/_{47}$ * $^{42}/_{46}$ * $^{41}/_{45}$ * $^{4}/_{44}$ = HYPGEOM.DIST(1,9,4,52,0) /9 =

0.0455850 (5%). iii) In this problem similar objects (i.e. other '9s')

can be in previous positions. P('9' on 9th card, irrespective of the

identity of cards 1 to 8) = 'one '9' at 9th no others in 1-8' OR 'one at

9th and one other in 1-8' OR 'one at 9th and two in 1-8' OR 'one at

9th and three in 1-8' = (HYPGEOM.DIST(1,9,4,52,0) /9) +

(HYPGEOM.DIST(1,8,4,52,0)* 3/44) +

(HYPGEOM.DIST(2,8,4,52,0)* 2/44) +

(HYPGEOM.DIST(3,8,4,52,0)* 1/44) = 0.0769231 (8%).

50. i) This can be answered by calculating the probabilities for '2 spades

by 2' = 13/52 * 12/51, 'by 3' as '1 and 3' (= 13/52 * 39/51 * 12/50)

OR '2 and 3' (=39/52 * 13/51 * 12/50), 'by 4' as ... 'by 41' as... .

Much less onerous is the use of HYPGEOM.DIST( ) in a similar

style to the above and 'by two' = HYPGEOM.DIST(2,2,13,52,0),

'by 3' = HYPGEOM.DIST(1,2,13,52,0) * 12/50, ... In Excel, put in

a column of numbers 2-41 for this, then reference them in the latter

formula all the way down, e.g., if '3' is in cell B4, P =

HYPGEOM.DIST(1,B4 -1,13,52,0) * 12/(52 - B4 +1), etc.

Multiply the probabilities by the card location and sum (use

SUMPRODUCT( )) to get E(X) '2 spades' = 7.571429 (about 8

cards on average). ii) If you look at your listing, you should see that

the highest probability is found with 5 cards (P = 0.109712 (11%)).

Index

BOOK ONE (separate volume)
Probability Basics

Available in Kindle and print

Book One: Probability Basics introduces the subject of probability. The nature of probability and how it is calculated are explored in a simple and understandable way. The concept of events and sample spaces along with counting methods bring the reader to probability distributions and simulation. Each section is illustrated by examples with randomising devices of common games and random influences in sport type games. These are worked out for all stages and spreadsheet formulae and functions make light work of difficult problems. Simulation methods (with *Excel® (2010) for Windows ®*, no macros) allow the reader to more easily solve math-intensive game problems, play roulette, roll different sized dice, enact sport streaks and deal a deck of cards and count the outcomes. Book One lays the basis for more detailed chapters on probability with randomizing devices and within specific games and sports (*Book Two*).